How to Think

How to Think

Understanding the Way We Decide, Remember
and Make Sense of the World

ROBINSON

ROBINSON

First published in Great Britain in 2021 by Robinson

5 7 9 10 8 6

A CIP catalogue record for this book
is available from the British Library.

ISBN: 978-1-47214-303-7

Typeset in Adobe Jenson Pro by SX Composing DTP, Rayleigh, Essex
Printed and bound in Great Britain by Clays Ltd, Elcograf S.p.A.

Papers used by Robinson are from well-managed forests
and other responsible sources.

Robinson
An imprint of
Little, Brown Book Group
Carmelite House
50 Victoria Embankment
London EC4Y 0DZ

An Hachette UK Company
www.hachette.co.uk

www.littlebrown.co.uk

Contents

Acknowledgments

The acknowledgments section appears first but is often written last. It's unusual in that it tries to find the fine lines between generous, genuine and gratuitous. You need to be complete yet concise.

There are many to acknowledge and thank in this section. Andrew McAleer provided guidance and support as editor from my initial book proposal to the final draft. His commitment to the publication process provided steady support and motivation. I never felt rushed or behind. It was a pleasure to write this way. And Andrew is a great editor. His ideas, suggestions and insights improved this book and improved my writing in general.

I also want to acknowledge my students and colleagues at Western University. I've been teaching and lecturing about psychology and thinking since 2003 and so many of the ideas I've written about here are the result of my efforts to teach about these topics, to make them interesting, to make them accessible, and to make them enjoyable. Heartfelt thanks to every student in every class I've ever taught. And thanks to my department colleagues, for the supportive and intellectually stimulating environment that we've all created at Western.

Locally, I want to thank Las Chicas del Cafe coffee, a small, family-owned coffee roaster from St Thomas Ontario. Their coffee is from family operated farms in Nicaragua and it's roasted perfectly. We've been fans of their coffee for years, and there's probably no way I would have completed this book without a steady supply.

My family is most deserving of my acknowledgment and thanks. My wife, Elizabeth and our daughters Natalie and Sylvie have been the most fun and loving friends and companions a person could ask for. We've inspired each other in so many ways and there's no amount of thanks that could be enough. And finally, a nod to Peppermint, our cat, who has appeared in several anecdotes in this book and others. Peppermint has been a near-constant companion during the writing process, often sitting on my lap as I type or read or on the corner of my desk while I work. She stares at the screen, not understanding a single thing I've written, but I love her anyway.

CHAPTER 1

An Introduction: About This Book

In 2003, I accepted a position as an Assistant Professor at the University of Western Ontario. When I arrived for work in July, I met with my department chair, James Olson, and talked about what courses I would teach. As a newly hired faculty member on the tenure track, I was entitled to a teaching break in my first year and was only required to teach one course. What should I teach? There was a course listed in the course catalogue that had been offered intermittently and was usually taught by an adjunct instructor or grad student. It was called 'The Psychology of Thinking' and was about reasoning, decision-making and other kinds of complex cognition. This seemed interesting.

I had been hired on the basis of my expertise in the psychology of concepts and this seemed like a good fit for me. So, I decided to make this course my own. I knew a lot about concepts and categories, and I knew some things about memory and decision-making but not as much about some of the other topics. I had to prepare a lot of the other material the same way that many other new instructors prepare. I read a lot of books and a lot of papers about higher-order, complex cognition. I prepared lecture notes and overhead transparencies. For those who do not remember, the overhead projector was a device with a mirror, a lamp and a lens that would take clear plastic transparencies (they were like clear paper) and project them on a screen for people to see. The overhead transparency was kind of like a handmade PowerPoint presentation. We use PowerPoint to project things digitally now, of course, but it's the same idea. You could actually write on the transparencies in class, which was a benefit, but you could also drop them on the floor, and they might fall out of order, which was not a benefit.

And so, in 2004, I took my position at the front of a small lecture room in the Social Science centre at Western, with my little stack of notes and a slightly bigger stack of overhead transparencies and I prepared to teach my first course as a professor. It was . . . all right. I probably stumbled on a few things here and there. I am sure that there were times that my lectures could have been better, but the students enjoyed the class and we all learned interesting things about cognition. One student wrote in my course evaluations at the end of the first year that the course was 'not as boring as I thought it would be'. That was all the motivation I needed to hear.

I have been teaching this course ever since then, over fifteen years. Although I also run an active research programme at my university, I derive considerable satisfaction from teaching, and in teaching this course in particular. It has grown from a single course of thirty students to three sections (two in class and one online) with a hundred + students in each class. Each week for most months of the year (because I teach this in the Summer sometimes), I think about how to bring together new and classic research in cognitive psychology and what that research has to say about what it means to think, to behave, to act and to be human.

Because I spend so much time thinking about the topic of thinking and thinking about how to make the research and concepts accessible and enjoyable, I decided to write a textbook for my course and other courses like it. This book, predictably titled 'The Psychology of Thinking', was originally published in 2015 and a second edition was published in 2020. That book was written for undergraduate students and it is properly a textbook that is designed to accompany a university or college course. The book presumed a certain level of familiarity with psychology. It is a good book and has been popular with students. But I also wanted to write a book that could be available outside the textbook market. A book that people might pick up and read during their summer vacation. Or a book that people might pick up at the airport and read on an airplane. A book that would find a home in the hands of people who are curious about how the mind works, how the brain works, how psychologists have studied the mind and brain, and how technology has allowed for new insights and discoveries about human behaviour.

The main goal of this book is to give you an enjoyable overview of how humans think. Are you interested in knowing how the different areas of your brain work, how objects in the world are recognised, and how you use your memory to guide your behaviours? I discuss those topics in this book. Are you interested in how people learn to use reason to make predictions and inferences about things? That's also in this book. I cover the history of cognitive psychology, perception, language, decision making, cognitive errors and the brain. I cover how we use language to guide our behaviour and to influence the behaviour of others. Frankly, I try to cover as much as I can, and still I think I can barely scratch the surface.

However, I also wanted to make two broader points in this book. First, I've found that we are all inherently interested in how and why people (including ourselves) behave. We are social/cognitive creatures. We want to understand the behaviour of ourselves and others. We want to know why people do the things they do, and we want to be able to predict what they might do next. But in order to understand how and why people behave the way they do you need to understand how they think. And in order to understand how people think, you need to understand something about cognitive psychology.

I also wanted to make a second point, and this one is a bit more subtle. We make mistakes all the time. We forget things, misremember things, overgeneralise, stereotype, and sometimes even fail to see things that are directly in front of us. We also fall prey to all manner of cognitive biases and illusion, in which we display overconfidence, failure to consider contradictory evidence, and a tendency to base judgements on the first things that come to mind. We're often bothered by these errors, but they are often a by-product of a system that is actually working well. The cognitive architecture that allows us to reach decisions quickly, to make mostly accurate judgements, and to make intelligent, adaptive decisions works really well for us. But the same aspects of our cognitive architecture that help us to think quickly also sometimes produce errors. The errors may be annoying, and we can learn to reduce them, but they are still a product of how the mind works. Memory errors come about from the same processes that allow us to learn and generalise. Judgement errors come from the same processes that allow us to make quick, accurate assessments. We probably can't eliminate cognitive errors and biases, but we can learn to recognise them and reduce their impact. That is, we all fall into cognitive traps from time to time.

One of the best ways to avoid falling into a trap is to know where the trap is, what the trap is, and why it's there. Errors are just a part of who we are, they make us human, they are mostly unavoidable, and arise from the same cognitive processes that enable us to learn, think, and remember. That's the other main theme of this book.

The Importance of Knowing

It's useful and powerful to know how something works. The cliché that 'knowledge is power' may be a common and overused expression but that does not mean it is inaccurate. Let me illustrate this idea with a story from a different area. I'm going to use this rhetorical device often in this book, by the way. I often try to illustrate one idea with an analogy from another area. It's probably a result of being a professor and lecturer for so many years. I try to show the connection between concepts and different examples. It can be helpful and can aid understanding. Later in the book, I'll discuss how and why analogies work this way. My reliance on making analogies can also be an annoying habit.

My analogy has to do with a dishwasher appliance. I remember the first time I figured out how to repair the dishwasher in my kitchen. It's kind of a mystery how the dishwasher even works, because you never see it working. You just load the dishes, add the detergent, close the door and start the machine. The dishwasher runs its cycle out of direct view and when the washing cycle is finished, clean dishes emerge. In other words, there's an input, some internal state where something happens, and an output. We know *what* happens, but not exactly *how* it happens. We usually study psychology and cognition in the same way. We can know a lot about what's going in (perception) and what's coming out (behaviour). We don't know as much about what's going on inside because we can't directly observe it. But we can make inferences about what's happening based on the function.

So, let's use this idea for bit. Let's call it the 'dishwasher metaphor'. The dishwasher metaphor assumes that we can observe the inputs and outputs of psychological processes, but not their internal states. We can make guesses about how the dishwasher achieves its primary function of creating clean dishes based on what we can observe about the input and output. We can also make guesses by taking a look at a dishwasher that is not running and examining the parts. Or we can observe what happens when it is not operating properly. And we can even make guesses about the dishwasher's functions by experimenting with changing the input, changing how we load the dishes for example, and observing how that might affect the outputs. But most of this is careful, systematic guessing. We can't actually observe the internal behaviour of the dishwasher. It's mostly hidden from our view, impenetrable. In order to understand how it works, you would need to devise clever ways to infer what's going on. Psychological science turns out to be a lot like trying to figure out how the dishwasher works. It often involves careful, systematic guessing. But if done carefully enough, it allows for reliable scientific inferences to be made.

The dishwasher in my house was a pretty standard mid 2000s model. It worked really well for years, but at some point, I started to notice that the dishes weren't getting as clean as they used to. Not knowing what else to do, I tried to clean it by running it empty. This didn't help. It seemed like water was not getting to the top rack. And indeed, if I opened it up while it was running, I could try to get an idea of what was going on. Opening stops the water but you can catch a glimpse of where the water is being sprayed. When I did this, I could observe that there was almost no water being sprayed out of the top sprayer arm. So now I had the beginnings of a theory of what was wrong, and I could begin testing hypotheses about how to fix it. What's more, this hypothesis testing also helped to enrich my understanding of how the dishwasher actually worked.

Like any good scientist, I consulted the literature: YouTube and do-it-yourself websites. They told me that several things can affect the ability of the water to circulate. The pump is one of them. The pump fills the unit with water and pushes the water around the unit at high enough velocity to wash the dishes. If the pump is not operating correctly, the water is not able to be pushed around and does not clean the dishes. But that's not easy to service and also, if it was malfunctioning, the dishwasher would not be filling or draining at all. So, I reasoned that it must be something else.

There are other mechanisms and operations that could be failing and therefore restricting the water flow within the dishwasher. The most probable cause was that something was clogging the filter that is supposed to catch particles from entering the pump or drain. It turns out that there's a small wire screen underneath some of the sprayer arms. And attached to that is a small blade that can chop and macerate food particles to ensure that they don't clog the screen. But after a while, small particles can still build up around it and stop it from spinning, which stops the blades from chopping, which lets more food particles build up, which eventually restricts the flow of water, which means there's not enough pressure to force water to the top level, which means there's not enough water cleaning the dishes on the top, which leads the dishwasher to fail. Which is exactly what I had been observing. I was able to clean and service the chopper blade and screen and even installed a replacement.

Knowing how a dishwasher works allowed me to keep a closer eye on that part, cleaning it more often. It gave me some insight into how to get cleaner dishes. It saved me some money, too. Knowledge, in this case, was a powerful thing.

And that's the point that I'm trying to make with the dishwasher metaphor. We don't need to understand how it works to know that it's doing its job. We don't need to understand how it works to use it. And it's not easy to figure it out, since we can't observe the internal state. But knowing how it works and reading about how others have figured out how it works, can give you a deeper understanding of the true processes at work. And this can give you an insight into how you might improve the operation; how you can avoid getting dirty dishes.

This is just one example, of course, and it's just a simple metaphor, but it illustrates how we can study something we can't quite see. Sometimes knowing how something works can help in the operation and the use of that thing. More importantly, this metaphor can help to explain another theory of how we explain and study something. I am going to use this metaphor in a slightly different way and then I'll put the metaphor away. Just like I put away the clean dishes from the dishwasher. They are there in the cupboard, still retaining the effects of the cleaning process, ready to be brought back out again and used, their clean state a physical memory of the dishwashing process.

We can agree that there are different ways to clean dishes, different kinds of dishwashers, and different steps that you can take when washing the dishes. For washing dishes, we have three different levels that we can use to explain and study things. First there is a basic function of what we want to accomplish, the function of cleaning dishes. This is abstract and does not specify how it happens, just that it does happen. And because it's a function, we can think about it as computational in nature. We don't even need to have physical dishes to understand this function, just that we are taking some input (the dirty dishes), specifying an output (clean dishes), and describing the computations needed to transform the input into the output: namely that you need to remove the food waste and debris.

However, we can also describe the process with less abstraction and more specific details. For example, a dishwashing process should first rinse off the food, then use detergent to remove grease and oils, then rinse off the detergent, and then maybe dry the dishes. This is a specific series of steps that will accomplish the computation described earlier. It's not the only possible series of steps, but it's one that works. This is an algorithm. It's like a recipe. When you follow these steps, you will obtain the desired results. If we want to study dishwashing at this level, we focus attention on the steps. This level of analysis helps us understand the importance of each step and to examine what happens at each step with enough detail to build a simple model of the process, maybe drawing out each step, so that we could design a dishwasher simulation. This algorithmic level would probably be enough to create a dishwashing blueprint.

If we really wanted to get to the bottom of things, however, we might want to study dishwashing at an even more specific level. After all, there are many ways to build a system to carry out these steps in the algorithm so that they produce the desired computation. My specific dishwasher is one way to implement these steps. But another model of dishwasher might carry them out in a slightly different way. And the same steps could also be carried out by a completely different system (like one of my kids washing dishes by hand, for example). The function is the same (dirty dishes → clean dishes) and the steps are the same (rinse, wash, rinse again, dry) but the steps are implemented by different systems (one mechanical and the other biological).

We can study dishwashers and dishwashing systems at each of these levels, and each level places the emphasis on different aspects of the overall system. That is, dishwashing is one simple task but there are three ways to understand and explain it: a computational level, an algorithmic level, and a representational level. My dishwasher metaphor is pretty simple and kind of silly. But there are theorists who have discussed more seriously the different ways to know and explain psychology and the mind. Our behaviour is one, observable aspect of this picture. Just as the dishwasher makes clean dishes, we behave to make things happen in our world. That's a function.

And just like the dishwasher, there is more than one way to carry out a behavioural function, and there is more one way to build a system to carry out the behaviours. The late and brilliant vision scientist David Marr argued that when trying to understand behaviour, the mind and the brain, scientists can design explanations and theories at three levels. We refer to these as Marr's *Levels of Analysis* (Marr, 1982).

Marr worked on understanding vision. And vision, which I will discuss more in Chapter 4, is something that can be studied at three different levels. Marr described the *Computational Level* as an abstract level of analysis that examines the actual function of the process. We can study what vision does (like enabling navigation, identifying objects, even extracting regularly occurring features from the world) at this level and this might not need to be as concerned with the actual steps or biology of vision. But at Marr's *Algorithmic Level*, we look to identify the steps in the process. For example, if we want to study how objects are identified visually, we specify the initial extraction of edges, the way the edges and contours are combined, and then how these visual inputs to the system are related to knowledge. At this level, just as in the dishwasher metaphor, we are looking at a series of steps but have not specified how those steps might be implemented. That examination would be done at the *Implementation Level* where we would study the visual system's biological workings. And just like with the dishwasher metaphor, the same steps can be implemented by different systems (biological vision vs computer vision, for example). Marr's theory about how we explain things has been very influential in my thinking and in psychology in general. It gives us a way to know about something. When we study something at different levels of abstraction, it can lead to insights about biology, cognition and behaviour.

Knowing something about how your mind works, how your brain works, and how the brain and mind interact with the environment to generate behaviours can help you make better decisions and solve problems more effectively. Knowing something about how the brain and mind work can help you understand why some things are easy to remember and others are difficult. In short, if you want to understand why people – and you – behave a certain way, it helps to understand how they think. And if you want to understand how people think, it helps to understand the basic principles of cognitive psychology, cognitive science and cognitive neuroscience.

That's what this book is about.

About Me

In the book *Travels with Charley: In Search of America*, John Steinbeck makes the point that people don't really take a book too seriously until they know something about the person who's writing it. It's a lesson that he learned early on, before he became a well-known writer. I think it's a good point. It helps to provide some context and to personalise the reading experience. Although this book isn't about me per se, I'm writing from a certain perspective that is a result of my experiences. I'm writing from a point of view that I developed over many years. My background, however mundane it may seem, forms the intellectual soil from which these ideas spring and grow. I want to share a little bit of information about who I am, and why I am interested in cognition.

You probably do not need to know too much of my early childhood, but I want to write a bit about my academic background for context. I grew up in the 1970s and 1980s in southwestern Pennsylvania, not far from Pittsburgh. After high school, I got a bachelor's degree from a small school in north-eastern Ohio called Hiram College. Hiram is one of those old, small liberal arts schools that are today somewhat of a dying breed. They were once really common in the north-eastern part of the United States, though much less so in other parts of North America and the world. There were only about a thousand students there from first year to fourth year, so my cohort, the class of 1992, numbered in the hundreds. I studied psychology then, and early on was interested in the human behaviour aspect of psychology as opposed to the clinical or counselling aspect. I carried out an honours project on moral reasoning and how people use their concepts of morality to guide decisions. I don't know whatever happened to the project and my component wasn't done well enough to publish. But it got me thinking about questions of behaviour and how we use our memories and concepts to guide things, which eventually became the focus of my own research.

I didn't attend graduate school right away, I decided to work for a year to earn some money and to think about my place in the world. But after that year, I decided that I missed the academic setting of university. I decided that I wanted to learn more about human behaviour and how the mind works. And I suppose I could have done that on my own as part of a hobby, but I wanted formal training. I enrolled in a master's programme in experimental psychology at Bucknell University. While I was there, I worked in a lab that studied music cognition. I wasn't interested in music cognition per se, but the lab I was working in was interested in how people remember familiar tunes. And this furthered my interest in how people use their memories for familiar things to guide future decisions. But a little knowledge is never enough. Not more than a year into my master's programme I was convinced that in order to understand how people think and why they behave, I needed to be trained in a larger programme. I sought out a PhD.

I attended the University at Buffalo between 1995 and 2000. The University at Buffalo is the flagship campus of the great State University of New York system. It's a large, research-intensive university with a large, diverse psychology department and one of the oldest, interdisciplinary cognitive science programmes in the country. And while I was there, I worked in a lab that finally seemed to satisfy my intellectual curiosity. This was a cognitive science lab and our emphasis was on understanding how people learn new concepts. Later in this book I will write a little bit about some of the research I did as a doctoral student, and how I discovered that when people are learning categories that are made up of only a small number of things, they tend to rely on memories for those individual things. But when they are learning larger, more well differentiated categories, they tend to form abstractions. In particular, I found that when people learn about a concept from specific examples, they learn what is generally true about a concept first, before they learn specific examples. This seems counterintuitive and the result was interesting enough to become a well-cited publication.

After my PhD, I studied as a postdoctoral research fellow at the Beckman Institute of Advanced Science and Technology, which is at the University of Illinois in Urbana-Champaign. And since 2003, I have been a professor of Psychology at the University of Western Ontario (aka 'Western') which is one of the largest research universities in Canada. Although I've done a lot of other things and I've had a number of other jobs, it's pretty safe to say that I've been interested in the study of psychology – specifically cognitive psychology – for nearly the last thirty years. It's also safe to say that there's little chance of me losing interest in this topic. Like most people, I'm fascinated by how people behave. How they think. How they act. So much so that I decided to make a full-time career out of it.

At a large research university like Western, professors are usually involved in both teaching and research. As an academic, I teach and lecture on cognitive psychology at the undergraduate and graduate level. As a scientist and a researcher, I maintain an active research programme that involves training PhD students, master's students, and undergraduates. My research is primarily about trying to understand how the brain and mind work to learn new concepts and new categories and also how people use concepts to think, plan and make decisions.

But I'm not going to describe my own research too much in this book, because it's a small piece of the bigger picture. Instead, I'm going to be describing and explaining the science of cognition. I'm going to be describing what cognition is and how we can study it. I'm going to be describing how the brain works to create cognition. I'm going to be describing how thoughts, behaviours and actions can be explained by studying the brain and by studying cognition. I'm going to be explaining what psychologists and scientists know about cognition and how it can be both useful and interesting to know a bit more about how you think, how you behave, why some things are easy to remember and why other things are not so easy to remember.

I'm going to be writing about how you learn to read, why you never forget how to ride a bike, and why there's always a cost to multitasking no matter how good you think you are at doing two things at once. I'm going to be writing about how you make decisions and solve problems and how the same general processes that help you make good, fast decisions also lead you to make mistakes. And most importantly, I am going to be writing about how scientists study these topics and ideas and how they make discoveries about the mind, the brain and behaviour. In short, I hope to give you an insider's guide to how the mind works.

For the most part, I'll be writing about how the brain and mind work from the perspective of a cognitive psychologist. That is, a psychologist who studies the thinking process. But as I hinted above, there are basic principles of psychology, cognitive science, perceptual science and neuroscience that are important to consider. Let's cover how these different approaches to the science of the mind came to be. In order to do that, we need to take a look at the history of psychology as a science and look at some of the insights and ideas from psychology and philosophy from the twentieth century and earlier. It's an interesting history in its own right and critical for understanding psychology now.

CHAPTER 2

A History of Cognitive Psychology

Every book, every concept, every political opinion, and every idea that we discuss today has some backstory. New ideas exist in the context of older ideas. In Chapter 1, I wrote a bit about my own backstory and how I found myself interested in concepts and categories, human behaviour, and how I ended up teaching people about thinking. That was my own context. Knowing a bit about someone's backstory helps us to understand who they are. Similarly, knowing a bit about an idea's backstory and context helps to make the idea more accessible and more understandable. After all, we use our memory of the past to make judgements and decisions about the future. We use our memory and knowledge about context to help us organise and perceive the present. In this chapter, I write a bit about the backstory of cognitive psychology so that we can better understand the present theories and ideas in this book. The history of cognitive psychology is more interesting that you might think. It reaches back to before the era of the Enlightenment. It had a revolution, bitter disagreements, big personalities and, as new technology comes on-line to image the brain, the present discoveries will be the backstory for the state of the science in the future.

Let's look at what makes psychology a science, what questions psychology is equipped to answer, and how it came to be the modern science that it is.

Fields of Study

There have been many ways to try to understand how the mind works. The ideas in this book, and the research and teaching that I do in general, fall under the umbrella of cognitive psychology. Cognitive psychology is a traditional discipline. This means that there is some agreement within the field on what the important topics are, what is known, what is not known yet, and what you can study. Within that field of cognitive psychology are the study of memory, attention, perception, language and thinking. Cognitive psychology does not include the direct study of neurotransmitters, the psychological effects of bullying, and treatments for depression, which fall under other subdisciplines of psychology. If you go to most university and college websites, you will find a Psychology department that probably has courses on 'Cognitive Psychology' or maybe a whole research area on the topic. This is orderly and gives the impression that we have some sense of organisation and internal consistency in our field. It gives the impression that we agree on what we study.

If only it were that simple.

It turns out that it's not so easy to classify these areas of study. For one thing, many cognitive psychologists draw influence from (and, in turn, influence) other fields and disciplines. Some psychologists study the biological aspects of cognition and behaviour. Other psychologists study how psychological research can be applied to improve learning. Still others are interested in how to measure behaviour and cognition. There are cognitive psychologists working in business, trying to understand how we act, behave and react to new products. And at the same time, changes and advances in technology and ideas have created new fields that overlap with these older fields and also with each other. Many of those topics fall into other areas as well. For example, there are psychologists and vision scientists that specialise in the study of perception who don't consider what they study to be cognitive psychology per se. There are psychologists who study thinking, reasoning and decision-making who might better be described as behavioural economists. There are many psychologists who study the effects of motivation on cognition who would consider themselves to be social psychologists. And researchers who study intelligence and IQ, which would seem to be related to cognition, nearly always come from the tradition of measurement psychology, rather than cognitive psychology.

It should be pretty simple to draw boundaries around fields, but in reality, it's not simple at all. But for the purposes of this book, I have to draw some lines. I am going to focus on three, broadly defined fields: cognitive science, cognitive psychology, and cognitive neuroscience. These three fields are all interested in understanding what the brain and mind do, in understanding how the brain supports thinking and cognition and how that in turn affects and drives behaviour. These fields also map very loosely onto the three levels of analysis that I described earlier. The interdisciplinary field of cognitive science tends to get excited about the computational level of analysis. Not exclusively, of course, but the emphasis is there in part because, as we'll see in a few pages, cognitive science takes a high-level view of cognition from the perspective of many different traditional disciplines. The second, cognitive psychology, tends to emphasise the study of process and function; the way the algorithmic level of analysis does. And the third field, cognitive neuroscience, works to understand how cognition is implemented in the brain. Yes, these fields overlap, and these are not the only distinctions that exist. But I'll try to make the case that these are the best possible categories for our discussion and that they are, in many ways, descendants of the older fields that preceded the scientific study of psychology.

Where do I fit? Well, I was trained as a cognitive psychologist. That means I study how memory works, how the mind operates to make decisions, how we classify and categorise things, and how we pay attention to some things while we ignore other things. And I rely primarily on behavioural, lab-based experiments to do this and I will often complement the research with computational models that describe how the different processes and algorithms might work. Cognitive psychology is concerned with the human mind, though we don't always use the term 'the mind' all the time. Most of us are not clinically focused, so we're not necessarily studying the diagnosis and treatment of psychopathological behaviour, mental health, or mental illness.

But sometimes I describe my research as cognitive science, especially if I want to emphasise the connections to other fields and disciplines. Other times, I might describe my research as cognitive neuroscience because I might be exploring a brain-based explanation for some set of results or because I am using a brain-based technique. But I would usually not refer to my work as 'neuroscience'. Neuroscience is a very broad discipline with its own traditions and training that I am not part of. I would also not use the terms 'neuropsychology' or 'cognitive neuropsychology' because these fields, while similar, tend to be about the clinical applications of brain research. These scientists often work within clinical settings with patients who suffer from disorders of the brain.

What do all these terms really mean? Is there a difference? Are we all sort of studying the same thing but using different terms? They aren't really the same, though many people (like me) use more than one of these terms to describe what they do. To answer the question, we probably need to go back a few years, or maybe a few centuries.

Precursors to Psychology

Cognitive psychology, cognitive science, and cognitive neuroscience are relatively young disciplines. But that does not mean that the things they study (the brain, the mind and behaviour) were ignored by earlier scientists. Far from it. The older forms of inquiry were more introspective and more intuitive. Introspection means 'to look inward' and if you don't have any other techniques or tools to measure thinking and behaviour, self-examination is not a bad way to start. These early, introspective traditions generated interesting insights and ideas, but the introspective technique lacked scientific rigour. However, it's still worth taking a closer look at some of this early research because of the way it influenced what came next. It helps to define what questions we want to ask about how the mind works. Our modern science and modern understanding of the brain and mind carry vestiges of these early discoveries. People have been interested in thinking for as long as we have been thinking. But it's only been in the last hundred years that humans have been able to study thinking and cognition in our modern, scientific way.

What are the precursors to modern psychological science? I don't want to go too far back, so I will begin briefly with some of the European Enlightenment-era philosophers. John Locke, for example, was an English philosopher whose important work was done in the late seventeenth century. Locke was prolific, influential, and made contributions to political science, economics and philosophy. He was also one of the very first thinkers to put forth a thoroughly modern idea of how the mind works. Locke's fundamental contribution to psychology is the idea that knowledge is not innate. Humans are not born with ideas, thoughts, and concepts. Rather we must acquire what we know from direct, sensory experience with the world. He argued that the mind was a 'blank slate' at birth and he referred to this as the *'tabula rasa'*, which is a Latin phrase that means a blank slate or writing tablet. None of us write on slates, of course, so this common metaphor may not make much sense. But a writing slate would have just been a small, portable chalkboard that was at one time made with actual slate stone: like a Microsoft Surface Pro with only one feature and really low screen resolution. The same metaphor could be expressed in modern times by saying the mind is a 'blank page' or an 'empty sheet' or even a 'new file'.

We call Locke's idea 'Empiricism'. We are born not knowing anything about the world, but we do have sensory systems with some basic, fundamental constraints. Through experience and observation, we learn how the world works, how to speak and read, and how to acquire new knowledge. We learn by noticing associations between things, by observing the natural correlation between events, and by making inferences about cause. Locke's ideas of how we acquire new knowledge and then extend our knowledge to new situations were developed further by David Hume and his work on association and induction. I'll have a lot more to say about Hume and induction later in this book, but his primary contribution was to explain some constraints on the blank slate idea. While Locke argued that humans have an innate capacity to reason, Hume suggested that we have no such capacity. This presents a paradox: if we have no capacity to reason and generalise, how do we learn anything at all? That is, how do we even know to start writing on the blank slate? Hume claimed that we learn to make predictions, inferences and conclusions about the world by the process of induction. With induction, we rely on our past experiences to make predictions about the future and, according to Hume, we have an instinct or habit to do so. In other words, the mind is not a completely blank slate. It's a slate with some rules. A slate with a memory. A slate that can generalise from the past.

Although we now understand a lot more about the nature of the brain, the role of genetics and the constraints of our cognitive and sensory systems, our modern understanding of the mind and brain is still basically an Empiricist view. We take it for granted now, but it was not always the case. Prior to Hume and Locke, people were just as likely to assume that ideas, thoughts and concepts were innate and/or divine. This idea that ideas are innate goes well beyond the assumption that we inherit natural abilities from our parents. A true nativist view believes that the concepts and ideas themselves are already within, waiting to be discovered. The French philosopher who is considered to be a founder of the modern Enlightenment, René Descartes, believed that concepts and ideas are innate and with us at birth. His idea (I suppose he even believed that his ideas about ideas were innate) was that our soul, which is semi-distinct from the body, already has access to idealised knowledge directly from God and that we can learn to uncover these truths with time and reflection. Descartes' ideas were dualistic in nature. That is, he believed that the body and mind were connected but not the same. He believed that the mind was not entirely of the physical world and that it had connections to a divine world.

I did not understand this idea when I first encountered it as an undergraduate. It just seemed to make so little sense to me. But it did start to make sense when I took a look at the broader historical context. Descartes was born at the very end of the sixteenth century at a time when Europeans were experiencing the changes of the so-called 'Age of Exploration' and the Reformation. Descartes was Catholic and thinking about that makes it easier for me to imagine him struggling with a duality of a different kind. Trying to develop a modern understanding of the mind that still fit within the earlier, middle-ages framework in which God played a role in everything that happened. Descartes straddled the line. So, I can see his dualism as a natural outgrowth of that line. With one side of the line being the magical, metaphysical and divinely inspired ideas of the past and the other being rational, physical and Earthly inspired science of the future.

Although there is some intuitive appeal to the idea that thoughts come from within, we take the basic ideas of empiricism so much for granted now that it is not easy to grasp the Cartesian[1] concept of innate concepts. However, there are still aspects of this view that influence how modern

---

1 Cartesius is the Latinised form of Descartes' name. *Cartesian* is then the adjective version of his name for ideas and works associated with him. Cartesian coordinates, or X-Y space, is one example. Descartes was said to have had an insight, which he believed to be divine, when watching a fly on the ceiling. He realised he could describe the location by using the lines where the ceiling and wall met and that any location on the plane could be described with these two coordinates. Descartes and the fly, Newton and the apple, Archimedes in the tub, we love to tell stories about people discovering things via insight while sitting around, kind of spacing out. This is, evidently, what people did before the internet.

psychology tries to understand how we think. Our thoughts and ideas do seem come from within us. People have all sorts of inner thoughts and ideas. Furthermore, the basic neurobiology of how we perceive, attend and think seems to be partially determined by our genes and is a product of evolution. Although evolution and natural selection itself are the result of external, environmental pressures on the organism: our genes store a record of how our ancestors adapted to pressure. That is, even biological explanations are not innate in the Cartesian sense. But it still makes sense to study some aspects of cognition and thinking as an innate process.

But there's also an appeal to the idea that our ideas, thoughts and knowledge come from without. After all, we rely on our memory, which is a representation of things that happened, to plan actions and make decisions. We speak the language that we do because of who and what we are surrounded with. We see the world through the eyes of our culture. Everything we know comes from our experience with the world and we learn new things about our world through the concepts we have formed from these experiences.

In popular parlance, this tension between explaining our thoughts according to what cognitive abilities we are born with vs what we acquire is often referred to as the contrast of 'nature vs nurture'. According to this contrast, we imagine that our psychology is either a product of what nature has given us, the descendent of Cartesian nativism, or it is a product of how our minds have been nurtured, the descendent of empiricism. Despite this being framed as a contrast, nativism vs empiricism or nature vs nurture, no one really believes that this is an either/or distinction. Rather, the contrast suggests that both play a role in our mental development, in how we form concepts and ideas, and in our psychology. We acknowledge that some psychological processes and abilities are influenced more by genes and biological constraints and others are more influenced by what we experience. This is the dominant view in our modern understanding of the mind. Our minds are neurobiological sort-of blank slates that work in predictable ways, governed by principles that we have inherited from biology. Blank slates with rules, constraints, biases and principles. Cognitive psychologists try to understand the rules, constraints, biases and principles of how these slates operate.

For centuries and centuries, philosophers (including Descartes, Locke, Hume and others that I will discuss), clerics, physicians and thinkers have tried to understand where our thoughts come from. Although this earlier work was important and still shapes our world view today, it was not until the end of the nineteenth century that a few scientists began to apply the scientific method to try to answer the question of 'where do thoughts come from?'

The first serious attempt to do this was Wilhelm Wundt in the late 1800s. Wundt was a physician in Leipzig, Germany, who wanted to understand the processes of the mind in the same way that physiologists can study the structure of organs and systems in the body. The problem was, of course, you can observe things like blood flow, viscera, bone and fluid in other bodies, record what you see, and develop theories from that observation. You can't do that with thinking (we can sort of do this now with neuroimaging, but that's a recent development and I'll talk about this later in Chapter 3). Wundt realised that if he was going to be serious about the scientific study of thought, he needed some way to measure and record what was observed. Good measurement and recordkeeping are essential to science. Science without measurement and recordkeeping is just speculation and fiction. As the Industrial Revolution gave way to the modern twentieth century, scientists of the mind began to look for ways to quantify and measure. And with Wundt's work, we come to the beginnings of psychology as a science.

The Beginnings of Experimental Psychology

'Psychology' is a very broad term. In popular use, it can mean many things. The most common definition is clinical psychology. We think of a psychologist as someone who interacts with clients or patients to help them with their mental health and their well-being. And that's certainly an important job description for psychologist. But there are other kinds of psychologists. Outside of the clinical domain, we can refer to this as 'experimental psychology' (or sometimes 'psychological science' though that also includes clinical work, too). Experimental psychology is best defined as the application of the scientific method to the understanding of human behaviour. One of the most important things in the scientific method is measurement. Scientists want to be able to measure things, whether they're measuring atoms, spores, body mass, atmospheric pressure or human behaviour, there has to be an agreed-upon way to measure things. The things that you measure and the way that you measure them, will define the kinds of data that you can collect from the world. This in turn affects the things that you can study, the questions you can ask, and the conclusions you can draw from your research. In a very real way, science is driven by the precision and limitations of the technology for measuring and recording things.

Early in the history of experimental psychology, however, there were no agreed-upon standards. No one had ever done anything like this before. No one had ever used the scientific method to study behaviour. But in the late nineteenth century, some researchers – taking their cue from medicine, physiology and biology – began to develop ways to observe and analyse behaviour. Wundt was the foremost of this group. He was interested in understanding how people created and understood perceptual experiences. For example, suppose someone asked you to choose a red card from an array of four different coloured cards. What goes on in your mind as you make that decision? Seems simple enough, just pick the red card. To carry out the very simple task of picking up a red card from an array of four coloured cards, you need to be able to cause your eyes and hands to respond to a verbal statement by working in conjunction to carry out these behaviours. Take a moment to consider:

- You need to hear the instruction.

- You need to understand the instruction.

- You need to direct your eyes to the cards.

- You need to direct your attention to each card.

- You need to recognise the different colours, perhaps by comparing them with some memory or internal representation.

- You need to make a decision about which card is the best match.

- You need a decision criterion (i.e. how close should the match be?).

- You need to make your hand reach toward the red card.

And that is nowhere near a complete list; even the first statement, 'you need to hear the instruction' assumes additional information about auditory perception and speech recognition. Each step involves sub-steps and subroutines. Picking up a card in response to a verbal request involves a long list of steps. And yet most of us would be able to do this so fast and so effortlessly that it would be difficult to describe. How would you ever measure all of these steps let alone just describe them?

In the absence of any other kind of measurement technique, Wundt developed something called 'trained introspection', which means looking inward. Experimenters in Wundt's lab would concentrate on observing their own thoughts and behaviours. Unlike naïve introspection, in which we might be dimly aware of what transpires in the mind, trained introspection requires concentration and considerable practice to be able to generate internal consistency in observation.

So, when you are asked to choose a red card from a pile of four different coloured cards, you might first concentrate on hearing the instructions and observing that one of those words reminds you of a colour. You may then observe your eyes being drawn almost automatically to the cards in front of you and briefly scanning to look for the red card. But how do you know which one is the red card? You need to introspect a little bit more to think about how you recognise colour for what it is. This takes time, practice and effort.

Wundt, and later his student Edward Titchener, developed what we now call *structuralism*, because he was interested in uncovering the structure of thought. At the time, there wasn't a clear consensus that all of these thoughts happened in different areas of the brain. Structuralism is not concerned with brain structure but rather with thought structure. Structuralists were trained in introspection just like we might train physiologists in basic anatomy or chemists in the ability to measure with a pipette.

If you've done any reading on meditation and mindfulness, this might remind you of the kind of exercises that people engage in when they are first learning to be mindful of their own thoughts. Introspection, like mindfulness, involves training yourself to be aware of what is happening in your mind. It can provide great insights into the complexity of perception, memory and thinking. Unfortunately, introspection is not very scientific or reliable. Researchers quickly realised that introspection as a technique is unreliable across different research labs. It also tends to ignore many unconscious influences. For example, if you were asked to choose a red card from a pile of four different coloured cards, your hand is likely to start moving towards the red card as soon as you hear the word 'red'. It is very difficult to introspect on that aspect of behaviour. Many visually guided motor actions are involuntary. Your eye movements towards the red card would also be involuntary and unavailable for inspection. As a result, introspection is inappropriate for many of the basic aspects of human cognition and behaviour. We just can't think objectively about how we deploy our attention, how we recognise objects, and how we retrieve things from memory. We are usually aware of the by-products, the memories and the contents of our thoughts, but we are not aware of the cognitive and neural processes that provide those by-products.

Behaviourism

Wundt's work, along with Titchener's work, was important, but incomplete. Introspection is just not the right technique for understanding how our minds work. It's too variable, too difficult to control, and too limited to be a measurement standard. Psychologists tried to get their act together and to develop systematic ways to measure behaviours with precision and objectivity. We usually refer to the second phase as *behaviourism* because these psychologists pushed back against Wundt (and also the psychoanalysts like Freud) who were trying to study these internal, subjective mental states. Instead, they argued that for psychology to be taken seriously as a science, it should limit its focus to only those things that can be measured and observed objectively. As a result, psychologists like John Watson and B. F. Skinner began to study behaviour as a function of stimulus inputs (what an organism can see or hear) and behavioural outputs (what an organism does in reaction to the stimulus).

The clichéd psychology experiment from this era is a rat in a cage, pressing a button or lever in response to a signal, and receiving a food reward. Behaviourists chose to study animals like rats and pigeons not so much because they were interested in what rats and pigeons could do, but because rats were a convenient model for learning in general. The assumption is that all organisms would follow the same basic principles. And for associative learning, behaviourists were right: rats, cats, monkeys and humans all display some of the same patterns. For example, my cat Pep (short for Peppermint) has learned to wake me up in the morning by carrying out a long and complex series of actions. Pep will meow, pick the wardrobe door open, rattle the window blinds, flop around in the bed, and sometimes even slam the door. She does all of this between 5:00 and 5:30 *prior* to my morning alarm. Her brain has linked together some association between the behaviours she carries out because they predict my waking up (and delivering food). She's learned something about the correlation between the set of behaviours and the eventual response. In Pep's tiny little cat-mind, she's made what amounts to a causal inference about her behaviours causing some of my behaviours. She does not know or understand why they fit together, only that they do fit together. She has learned these associations through a process that behaviourists call *operant conditioning*. The behaviours she does; scratching, meowing and picking things, are all part of what cats normally do. But if I wake up and feed her, she will learn to associate one or more of those behaviours with the food. If I don't wake up and feed her, she will not learn to associate the behaviours. The learning is gradual but not involuntary. She is motivated to learn. Pep really wants to wake me up because she really wants to be fed.

This gradual shaping of behaviour was first discovered by an early psychologist named Edward Thorndike. Thorndike, like many early psychologists, was intensely interested in mental phenomena but not quite sure how to study them. At first, he tried to study mental telepathy in children. As you can imagine, this was a failure, primarily because telepathy is impossible. But he noticed that the children he studied could pick up on very subtle cues, like the experimenter's involuntary and unconscious movements, and that could *seem* like telepathy. Good poker players, for example, can read subtle cues or 'tells' like involuntary facial movements. This led him to try to understand how behaviour might be shaped by reinforcements in the environment. After some less than successful attempts to study the intelligence of chicks (he was not allowed to have chickens in his apartment) he developed an apparatus called the 'puzzle box' for studying cats. The cat was locked in a box that had all kinds of levers and pulls and they would try all sorts of things to get out. Thorndike designed the box so that a specific sequence would unlock the door and the cat could get out and eat. This was basically an escape room for cats, but rather than having fun with friends, like humans do, the cat was probably either mildly ter-rified or mildly bored. Once they discovered the sequence, they would be trained and tested again, and Thorndike found that their behaviour was shaped by the eventual reward. This seems really obvious to us now, but at the time was ground-breaking because it demonstrated that cats could retain a memory trace of what they had just done, and that trace could be strengthened (or 'stamped' in Thorndike's terms). And just like evolution and natural selection can select for and favour certain traits that help the organism survive in the long run, this shaping behaviour can select for and favour behaviours that help the organism survive in the short run.

Humans do the exact same thing, of course. I notice a great example when I park on my university campus. Students, faculty and staff who buy a parking pass have a hang tag for their vehicle that is read by a radio sensor (RFID) that will open the parking gate. You may have one of these if you work at a university, hospital or a large office in a city. But no one really understands how it works. I see people drive up, try to position their cars in just the right place, try to position their hang tag in just the right place, pull up slowly, stop, move, stop, and just like my cat trying to go through this whole sequence of behaviours to predict my waking up, the drivers go through a whole sequence of sometimes unnecessary actions to get the gate to open up. We don't know which actions are necessary to cause the gate to open up, only that some sequence of them is. The gate usually opens and whatever we were doing prior to that is reinforced. If you just happened to be driving back and forth when it opened, that would be reinforced, even if the whole sequence was not required to open the door. The association is gradual. Things that seem to work are reinforced. Things that do not work are not reinforced. And some irrelevant behaviours might end up being reinforced, because they co-occur with the outcome. We sometimes call these 'superstitions'.

But here's the biggest difference. If we wanted to, we could try to understand how the gate works. We could use our ability to reason. We could use our language to state a hypothesis and test it. We could read an article online[2] and try to understand how the RFID system works, what its

---

2 When I went to search online for RFID parking, I was confronted with page after page of 'how to hack parking'. It seems like a really common denominator for humans is how to beat the system and take shortcuts.

range is, and we could adjust our behaviour. This takes planning, language and a theory of mind that assumes that other people know how things work and are giving us reliable information. And when this is implemented, it's not as gradual. If you read the instructions, carry out those actions, and the gate opens, you will just keep doing it. In other words, we have the luxury of being able to understand the cause and effect. Cats don't have that luxury. We have that luxury because we have language. And behaviourism, as an approach to psychology, turns out to be ill-suited to explain how and why we use language.

Cognitive Psychology

In 1957, the behaviourist psychologist B. F. Skinner published a book called *Verbal Behavior* (Skinner, 1957). And by 1957, Skinner had already become the most famous psychologist in the world. He was prolific, interesting, well-known and a somewhat notorious public figure. As a graduate student, he had invented the operant conditioning chamber (a.k.a. the 'Skinner Box'), which was an apparatus that tracks behavioural responses made by rats in the presence of different cues and reinforcers. It was even believed for a time that he had raised his daughter in an operant chamber (he did not). He was in many ways *the* authority on psychology by the mid-1950s. In Skinner's view, all behaviour could be accounted for by the fundamental mechanisms of reinforcement learning. And so, *Verbal Behavior* described a theory for how that might work, arguing that we learn to communicate because we're reinforced for saying some things and not others. That is, humans use language to obtain the things they need. When a child points to a toy or food and verbalises something, they get the toy or the food. With time, their verbal behaviour is shaped according to the same rules of reinforcement and the more general operant learning. If behaviourism psychology was to be a natural science of human behaviour, then the rules and laws of learning should generalise to the range of human behaviour, including language.

But the book was met with something of a backlash within the field of linguistics. Noam Chomsky, for example, wrote an extensive review and critique of Skinner's book. Yes, the same Chomsky who is now probably more famous for his debates with William F. Buckley, an appearance on *Da Ali G Show*, and his leftist politics. Chomsky's review pointed out that language is a complex behaviour that is learned with minimal examples and not much in the way of feedback (he later referred to this idea as the 'poverty of the stimulus'). Children say things that they have never heard and can be rewarded when they say something incorrect. When a toddler says 'I ah juice' the parent will get them some juice. The parent would probably not ignore them until they say, 'I'd like some juice'. Chomsky argued instead that there must be some innate behaviours that make language all but impossible *not* to learn.

Chomsky's review in 1959 was a hit. Not so much a hit with the general public, but a hit with many psychologists and linguists who viewed Skinner and other behaviourists as too rigid, too committed to a specific theory, and too dismissive of other approaches to study behaviour and the mind. Others viewed Chomsky's review the way you might view a critic who likes (or hates) the same movies you like (or hate). It provided some justification and a feeling of 'yes, thank you!' to people who had been studying phenomena outside the realm of behaviourism. Chomsky's review is now regarded as one of the founding documents of what came to be called 'the cognitive revolution'. This was not something that was planned out in secret or that was fought in lecture halls, but it was a rapid shift in thinking. And the notion that this shift in thinking was a 'revolution' became an origin story for experimental cognitive psychology.

Yes, cognitive psychology has its own origin story. According to the slightly fictionalised version, experimental psychology departments in the 1950s were nearly all committed to teaching behaviourism. But people wanted to break out of the confines of studying behaviour as a result of reinforcement contingencies, and (it was imagined that) Chomsky's review pulled back the curtain to show the limitations. It's an overstatement, of course, but it has become part of the story of experimental psychology. And indeed, the 1960s were an incredible time of discovery for experimental psychology. Realistic or not, the success of cognitive psychology in the wake of Chomsky's book review might have reinforced the idea that the review was the stimulus or the catalyst for the later development and strengthened the association between the review and the later discipline.

There was another, far more practical but less dramatic aspect to this so-called revolution. By the late 1950s and into the 1960s, most large research universities owned and operated mainframe computers for research. As these became more widespread, they became available for psychologists to use. The development of the digital computer made it possible to measure and analyse data collected from people that was well beyond the response rates that were measurable by Skinner's operant chambers. Psychologists were able to develop ways to measure reaction time (the millisecond-precise measure of how fast a person can respond to something). They were also able to develop ways to present words and pictures visually to research subjects with very precise timing. This had been possible before this period, but the techniques expanded in the 1960s. These technological advances pushed experimental psychology forward rapidly in ways that might not have been possible without computers.

But more important in my view is that the computer made possible a new metaphor for the mind. Remember that behaviourists developed the limiting approach that they did, in order to rein in the less scientific forms of measurement that were being used by the structuralists before them. Their model of the mind emphasised mechanical functions and computations too, but their approach did not allow for any observation or description of internal mental states. In doing so, they effectively limited psychology to the study of the inputs and outputs of behaviour. A computer has inputs and outputs, but it also has a clear and observable internal state. The electronic circuits and tubes of early computers made it easier to see that the external world (inputs) could be *represented* through connections of circuits. Different internal architecture can affect the performance of the systems. The order of information processing steps matters. This representation could influence the output of the system and is even something that could be studied and analysed. Unlike structuralism, which emphasised the thinking process, or behaviourism, which studied laws of behaviour, the study of internal representations as in a digital computer provided some way to study unobservable, internal states.

The development of the computer enabled what we might call the 'mind is a computer' metaphor and this metaphor came to influence how we think about the mind and brain and how we want to study it. Cognitive psychology, then, is the study of mental behaviour and mental representation. It is also the study of the mind, which is the name we often give to the information processing and computation that happens in the brain.

As we have seen in earlier sections, the metaphors of the mind are both a product of their time and also a driving force in how science is conducted. Descartes, in Renaissance-era Europe, saw the influence of God and the divine. For him, the mind was not altogether part of the body. It was also part of the divine. And so, nativism is a reasonable outgrowth of this metaphor: your mind is what God designed. The mind had been seen as a blank slate during the age of the Enlightenment, as functional anatomy in the Age of Darwin, or as a stimulus-response engine during the Industrial Revolution. These metaphors – the divine, the blank and the mechanical – frame scientific inquiry. The limitations of these metaphors also prompt shifts in scientific thinking. The computer metaphor is the metaphor that drives cognitive psychology. And even as we've learned more about the brain, down to the level of neuron, the metaphor seems to be sticking. It may be a much more profound paradigm shift.

Cognitive Science

The history of science and technology is often delineated by *paradigm shifts*. This is a fundamental change in how we view the world and our relationship with it. The big ones are sometimes even referred to as an age or a revolution. The Space Age is a perfect example. The middle of the twentieth century saw not only an incredible increase in public awareness of space and space travel, but many of the industrial and technical advances that we now take for granted were by-products of the Space Age. Space exploration enabled us to think differently about our planet and our place in the universe. For the first time in human history, it was possible for our species to take the long view; to be able to see our entire planet in one picture. To be able to see how we, as humans, are a small part of the entire system and not the centre of it. The Space Age and the race to fly manned space missions that reached orbit and eventually the Moon was made possible by developments in chemistry, computer science, materials science and communication. Many of the technological advances that we take for granted now were developed to support the space programme (and the military more generally). The smartphone that you keep in your pocket or in your hand is the descendent of materials developed by defence and government contractors. It relies on communication protocols developed for space exploration and defence, and it uses the GPS system that would never have been possible without the communications satellites that were originally launched in the 1950s. The internet itself, while not a direct outgrowth of the space programme, grew alongside it, as a computer communication system developed by the

US Department of Defense and influenced by the cold war. I could go on, but I think this is enough to make it clear that the Space Age, the middle of the twentieth century, brought about a paradigm shift in thinking and a technological shift that changed much of human life on Earth.

I believe we are at the beginning of a new age, and a new, profound paradigm shift. I think we're well into what I would call the Cognitive Science Age. I'm not sure anyone else calls it that, but I think that is what truly defines the current era. Other possible descriptions might be the Computer Age, or the Age of the Algorithm. Or maybe the Data Age. We have come to rely on computer and data to an extent never seen before and much of this arose in the twentieth century as cognitive psychology met computer science, linguistics and neuroscience and the term 'cognitive science' was born.

I also think that an understanding of cognitive science is essential for understanding our relationships with the world and with each other. I say this because in the twenty-first century, many of the computer-based approaches to understanding behaviour − artificial intelligence, machine learning and deep learning − are now being fully realised. Every day, computer algorithms are solving problems, making decisions and making accurate predictions about the future; about our future. Algorithms decide our behaviours in more ways that we realise. They decide what we read on social media; they decide what advertisements we will be exposed to. They are able to fine-tune these into 'advertainments' that capture our attention in ways that can be very difficult to resist, in part because of all the data we provide to media and advertising companies which are then analysed by complex and powerful computer algorithms.

As I am writing, I am listening to music on Spotify, which uses algorithms to analyse my listening preferences to fine-tune its service and keep me listening and paying. Netflix and Amazon Prime do the same with video. And of course, my phone helps me do all kinds of things by algorithm. Every time I use it to do something, I also contribute to the underlying data that makes their (Google, Apple, Facebook, etc.) algorithms even better. For example, when I search for restaurants or retail outlets on Google Maps, I get recommendations that are useful to me. The more I search, though, the more information I give to Google that they can use to improve the search. I also use a service called Google Lens on my Android phone (it's really amazing, by the way). I point my camera at an object I want to identify: usually a bird, a plant or an insect. When I activate Google Lens, it analyses features of the image and searches, via Google, the internet for matches to help me identify things. This a useful for me if I want to know what kind of butterfly or plant I'm looking at. But this information is also very useful to Google because it will help their algorithms to learn more about the natural world. For us, it's a search. For Google, it's input for training.

Algorithms run more than just media companies. We look forward to autonomous vehicles that will depend on the simultaneous operation of many computers and algorithms. Even non-autonomous vehicles have sensory systems and processing algorithms. When I bought a new car in 2019, I was amazed at progress in algorithm-assisted driving. This car is not an autonomous vehicle, but it can take over some things. With cruise control, it now is able to sense the speed of cars in front of me and slow down accordingly. It can warn about vehicles on the side and rear areas that would otherwise be in a blind spot. It can even warn me if it thinks my attention is wavering (like if I switch lanes without signalling). And with on-board GPS and/or a connected smartphone, it knows where I am all the time, and it can track data on speed, braking, accelerations, etc. This data helps me with driving but can also help the manufacturer design better algorithms. And my own driving behaviour has changed. The assistive technology is no longer novel. It's just a part of regular driving now.

Computers, machines and algorithms will (and have) become central to almost everything we do. And every time we rely on a computer algorithm to help us with something – to identify an object, to search for a location or to make a decision – we help to provide crucial training data to the algorithm. Every time we use a voice command via Siri, Google or Alexa, we provide more training data for those services. If the voice command was carried out successfully, then the algorithm learns from that. If it was not carried out successfully, then it learns from that as well. We are training these machines, these AI systems, these bots, to know what we need and predict what we *will* need. As they get better, we rely on them more and more, and in the process, we help them improve, which will then encourage us to rely on them even more, and so on. It's a reinforcement loop wherein the algorithms help us, and we help the algorithms. Is this exciting? Or is this frightening?

This is indeed a new age. This is indeed a paradigm shift. This is the culmination not only of cognitive psychology and computer science but of several related fields that are all broadly interested in the study of information processing. This is cognitive science. And I argue that our modern age that places data, algorithms and information as the primary raw materials and industries, should be called the Cognitive Science Age. As cognitive scientists (a label that I often self-apply) this new age is our idea, our modern Prometheus. Let's take a close look at cognitive science and how it brought together several different fields, how it developed, and why it matters.

Cognitive science is an interdisciplinary field that first emerged in the 1950s and 1960s (along with the digital computer) and sought to study cognition, or information processing, as its own area of study rather than as a strictly human psychological concept. As a new field, it drew from cognitive psychology, philosophy, linguistics, economics, computer science, neuroscience and anthropology. As the field developed, it took on a name, an approach and an academic society. Despite that, most scientists, even those that call themselves 'cognitive scientists', remain part of the established traditions. Some scientists even wonder if cognitive science still exists (Núñez et al., 2019). That's probably because cognitive science is an interdisciplinary approach rather than a field per se. Although people still tend to work and train in those more established traditional fields, it seems to me that society as a whole is in debt to the interdisciplinary nature of cognitive science. And although it is a very diverse field, the most important aspect in my view is the connection between biology, computation and behaviour. It is from these interactions that our modern age grew.

The influence of biology

A dominant force in modern life is the algorithm; a computational engine to process information and make predictions. Learning algorithms take in information, learn to make associations, make predictions from those associations, and then adapt and change. This is referred to as machine learning, but the key here is that the machines learn biologically.

For example, the algorithm (Hebbian Learning) that drives many artificial neural networks' learning was discovered by the psychologist and neuroscientist Donald Hebb at McGill University. Hebb's book on *The*

Organization of Behavior (Hebb, 1949) is one of the most important written in this field and explained how neurons learn associations. This concept was refined mathematically by the cognitive scientists Marvin Minsky, David Rumelhart, James McClelland, Geoff Hinton and many others. The advances we see now in machine learning and deep learning are an indirect result of cognitive scientists learning how to adapt and build computer algorithms to match algorithms already seen in neurobiology. This is a critical point: it's not just that computers can learn, but that the learning and adaptability of these systems can be grounded in an understanding of neuroscience. That's the advantage of an interdisciplinary approach.

As another example, the theoretical grounding for the AI revolution was developed by Allen Newell (a computer scientist) and Herbert Simon (an economist). Their work in the 1950s–1970s to understand human decision-making and problem-solving and how to model it mathematically provided a computational approach that was grounded in an understanding of human behaviour. Again, this is an advantage of the interdisciplinary approach afforded by cognitive science.

The influence of algorithms

Perhaps one of the most salient and immediately present ways to see the influence of cognitive science is in the algorithms that drive the many products that we use online. Google is many things, but at its heart, it is a search algorithm and a way to organise the knowledge in the world so that the information that a user needs can be found. The basic ideas of knowledge representation that underlie Google's categorisation of knowledge were explored early on by cognitive scientists like Eleanor Rosch and John Anderson in the 1970s and 1980s (I discuss this research later in Chapter 9).

Or consider Facebook. The company runs and designs a sophisticated algorithm that learns about what you value and makes suggestions about what you want to see more of. Or, maybe more accurately, it makes suggestions for what the algorithm predicts will help you to expand your Facebook network ... predictions for what will make you use Facebook more.

In both of these cases, Google and Facebook, the algorithms are learning to connect the information that they acquire from the user, from you, with the existing knowledge in the system to make predictions that are useful and adaptive for the users, so that the users will provide more information to the system, so that it can refine its algorithm and acquire more information, and so on. As the network grows, it seeks to become more adaptive, more effective and more knowledgeable. This is what your brain does, too. It causes you to engage in behaviour that seeks information to refine its ability to predict and adapts. These networks and algorithms are societal minds; they serve the same role for society that our own network of neurons serves for our body. Indeed, they can change society. This is something that some people fear.

When tech CEOs and politicians worry about the dangers of AI, I think that idea is at the core of their worry. The idea that the algorithms to which we entrust increasingly more of our decision-making are altering our behaviour to serve the algorithm in the same way that our brain alters our behaviour to serve our own minds and body is something that strikes many as unsettling and unstoppable. I think these fears are founded and unavoidable, but like any new age or paradigm shift, we should continue to approach and understand this from scientific and humanist directions.

This is the legacy of cognitive science, and indeed the legacy of the development of experimental psychology from the nineteenth century onward. The breakthroughs of the twentieth and twenty-first centuries arose as a result of exploring learning algorithms in biology, the instantiation of those algorithms in increasingly more powerful computers, and the relationship of both of these concepts to behaviour. The technological improvements in computing and neuroscience have enabled these ideas to become a dominant force in the modern world. Fear of a future dominated by non-human algorithms and intelligence may be unavoidable at times, but an understanding of cognitive science is crucial to being able to survive and adapt.

I've traced the development of experimental psychology from the early nineteenth century structuralist days though behaviourism and cognitive science. But you may have noticed that I left something out: the brain! Although I think we're living in the Age of Cognitive Science, I think the real next frontier in this age is cognitive neuroscience, or the study of the brain and mind. I did not leave this out because I forgot to write about it or because it's not important. I left it out so that I can cover more in the next chapter. Let's discuss the brain.

CHAPTER 3

Understanding the Brain

When I was growing up, a friend of mine was in a tragic car accident. I don't remember all the details from thirty years ago, but I do remember that she was the only person driving and somehow on the way to school she lost control of her vehicle and hit a tree. The force of the impact broke the bones in her face and also damaged a significant part of the frontal area of her brain (the area known as the prefrontal cortex – sometimes referred to as the PFC – I'll discuss more about brain anatomy in a few pages). The damage was severe: not only were there facial and cranial fractures and damage to the brain from the force of the impact, but the neurosurgeons had to remove a small part of the front of her brain, a procedure known as a frontal lobectomy. She spent several weeks in a coma during the recovery.

My family and I visited her in the hospital a few times and for weeks, nothing changed. She was unresponsive. After a month or two, she woke up but with no capacity to speak and with minimal awareness of who her parents were. But even a damaged brain is able to heal itself, and in young people this regeneration can restore some functioning.[3] And so, with time, she regained some of her ability to speak, to listen, to walk around, to read, and eventually to carry on conversations with friends and family. It seemed like she would recover.

---

3 *Neuroplasticity* is a term that refers to developmental flexibility in the brain in general and is often used to mean the ability of the brain to adapt to damage and change, especially at early ages.

She missed her graduation that year and I went to university a few hours away, and lost touch with her and her family. There were no smartphones in those days and no social media. The sparse internet was for research universities and although I had an email address in the late 1980s, no one else did and so I never received email. The only way to communicate with friends and family was though letters. Letters were like emails or DMs, but you wrote them out by hand on slips of paper and sent those slips of paper in envelopes to the recipient who would then read your thoughts a few days later (the whole thing seems so strange now) or by telephone. And even then, we only had one payphone in my hall of residence, so it tended to be reserved for calls home. This is all a way of saying that I had no practical way to keep up even if I had wanted to.

A few years later, having recovered from the accident, she was finally able to graduate from her high school, and my family and I attended a reception afterward. Since I had been out of touch, I had no idea that she had recovered. I was eager to catch up with my old friend. We talked about high school, about her plans for the future, and also her accident. She looked almost the same, and sounded the same too, which was encouraging since the last time I saw her she could barely speak.

But here's the thing: *she was no longer the same person*. I don't mean this in a metaphorical way. I mean she really did not seem to be the same person at all. It was a like a new person was inhabiting the same body. Even as her physical appearance returned to a state that was close to her pre-accident life, her personality had changed noticeably. It was easy to look past the physical scars on her face but not easy to look past the changes in thinking, behaviour and personality. We accept that people change on the outside all the time. We do not accept that they change on the inside. Her aspirations had been to attend an Ivy League college and to make a difference in society in her professional choices. Her personality was what you would describe as studious, friendly, agreeable, and she was high on empathy. She thought a lot about others and was very close to friends. She was someone in whom you could trust.

After the accident she became difficult to understand. Initially, it took many months for her speech to return. And when it did, it returned slowly. She often confabulated and told stories that were sometimes confusing, sometimes funny, sometimes contradictory. She described events that were either complete fiction or were things that should be confidential. She made up things that were objectively untrue, such as what movie she saw earlier that week and often told conflicting stories about the same events to different people. When compared to how she used to be, she also lacked social inhibition: she was often unable to keep from blurting out inappropriate comments. Whereas her demeanour before the accident was that of a careful, intelligent, trustworthy and ambitious person, her demeanour after the accident was that of a person with a chaotic mental life. I should point out that these observations were from the year or two after her recovery, many decades ago. I don't know what has happened to her since then. Maybe she's recovered more fully and gone on to live a fulfilling life. I certainly hope she has.

I am beginning with this story because it lets me illustrate a few points. For one, this was the first time I came face-to-face with the connection between the brain and behaviour. I knew that thinking, memory and behaviours were brain-based functions, of course. And by then I had taken a course on *Introduction to Psychology*, so I had read about famous case studies. But I had never seen the connections between brain, mind and behaviour so clearly and directly before. Here was someone I knew, who looked and sounded the same as they did a few years earlier but acted and behaved very differently. The only change was the damage and subsequent removal of her prefrontal cortex.

But more relevant to this book is the way in which this case shows that specific behaviours can be tied to specific brain regions. That is, her prefrontal cortex seemed to control some aspects of her personality and specifically her ability to plan what she wanted to do and say and her ability to inhibit inappropriate behaviours. Cognitive neuroscientists refer to this idea as *localisation of function*. Although complex behaviours and thinking depend on many areas of the brain, some components of behaviours can be localised to specific areas of the cortex. As we'll see later in this chapter, there are areas of the brain specialised for understanding and producing spoken language, for recognising and processing faces, for coordinating your hand and eye movements, and (in this case), executing and inhibiting complex behaviours.

This example also allows me to highlight how the field of cognitive neuroscience came into being. Much of what we now know and understand about the connection between the brain, mind and behaviour was initially discovered through work with patients who suffered brain damage as a result of stroke, blunt force trauma or as a side effect for surgery to treat something else. Later in this chapter, I'll return to this case and discuss why the damage that she sustained affected her in the way it did.

This book is primarily about thinking and the mind. But we need to begin by understanding the brain; the organ that drives cognitive processing. Later in this chapter, I will cover a basic introduction to the brain's structure and to the methods that are being used in the study of cognitive neuroscience. I will discuss the long and interesting history of this field including some of the most famous and fascinating case studies that promoted an early understanding of how brain structures map onto aspects of personality and cognitive functions. And because research in cognitive neuroscience is some of the most influential and exciting research being done today, I'll also write about some fascinating research being done at universities and research institutes around the world, including some truly ground-breaking work being done at my own university on the nature of consciousness and the illuminating work by social psychologists who try to understand how the brain guides our political beliefs and social behaviours.

The Structure of the Brain

Let's begin with a quick tour of the brain. If our goal is to understand why and how people behave the way they do, then we need to understand more about cognition and information processing. And to do that, we need to understand how the brain works.

Inside your head, right beneath your skull, your brain is a dense organ made of protein and fat. It's shielded from direct contact with the outside world, but the brain connects to the world and extends your awareness through your eyes, ears, nose, fingers and other sensory systems. These inputs are connected to other inputs and those connections constitute your experience with the world. Those connections also represent things that already happened: your memories and your knowledge. Everything you think about, consider, decide, remember and consciously experience, happens in this organ. Simply put, your brain is who you think you are.

Despite what you may often hear in popular discussion, it's not true that the average person only uses 10 per cent of their brain. You use all of your brain all of the time. I'm not sure where this neuromyth[4] originally came from, but it's kind of absurd on the surface. Let's look at this claim in more detail. The human cerebral cortex is one of the most highly evolved structures in mammalian physiology and it's funny to think that we would have evolved with a complex brain that is 90 per cent inoperable by default. We don't make claims like 'you only use about 10 per cent of your liver' or 'the average person only uses about 15 per cent of their skin at any given time' so why would we believe this about the brain? Furthermore, in those cases in which someone really is using less than 100 per cent of their brain, like someone with damage to part of their brain, the effects are often clearly noticeable. And yet, we still make reference to the 10 per cent of the brain claim.

It is probably more accurate to say that we are only consciously aware of a small proportion of the overall activity in the brain. But that's a cognitive limitation, not a physiological one. This limitation is one that probably evolved as an adaptive advantage for us as well. It would be impossible to be explicitly aware of every brain process as it happens. The world presents a constant array of changing sensory information, but much of the detail is not relevant to what we're doing or thinking. We're also largely unaware of the constant brain activity needed to keep us breathing, standing, perceiving and living. Of course, we're not going to be aware of all of everything! We need some way to prioritise what we can pay a lot of attention to, what we can pay a bit of attention to, and what happens automatically without attention and awareness. I will discuss these ideas in much more detail in Chapter 5. For now, the point I want to make is that we're using all of our brain all of the time, but because of the way our cognitive system has evolved, we are only aware of a small proportion of that activity. This limitation or bottleneck is one of the most fundamental aspects that governs how we think.

With that common neuromyth out of the way, let's get back to a discussion of the brain. Your brain is about the size of a large cauliflower. Or more correctly, the average human brain is between 1120 and 1230 cubic centimetres with some minor variation due to genetics, sex, nutrition

---

4 *Neuromyth* is a term often used by scientists to refer to very common but unsubstantiated and/or factually incorrect beliefs about the brain. Common neuromyths include the idea that we only use 10 per cent of our brain, that some people are 'right brained' and more artistic, while others are 'left brained' and more mathematical, and that there are brain-based 'learning styles' (Dekker, Lee, Howard-Jones & Jolles, 2012)

and other factors. The brain is far from a uniform mass. The outer portion, what you would see if you were looking at a picture of a brain or looking at an actual brain, is called the cortex. That's where most of the cognitive action is, and that is also the area that I will be discussing in the most detail. Other areas just below the cortex are important for creating your memories, and for understanding emotion. These subcortical structures are also important for our discussion because they also affect how we remember, think and respond to things in the environment. There are also several other structures, variously referred to as the midbrain and hindbrain, that help to maintain your basic functions: keep your heart beating, keep you breathing, and keep you alive. I won't discuss this too much more in this book, though these areas are still important (because they keep you alive).

The human cortex, in addition to varying in its internal structure, also varies between different people. Some people have larger brains, and some people have slightly smaller brains. Overall size in the human cortex is generally correlated with overall physical size but overall size is not strongly related to intelligence or behaviour. There's a small correlation but many other factors play a role in what we call intelligence. There are biological sex differences, too.[5] A large-scale study of human brains from the UK Biobank programme confirmed that male brains tend to be just slightly larger than female brains (Ritchie et al., 2018). Some of that variance is to be expected, given the general difference in physical size between human males and females. The researchers found substantial overlap between male and female brains, but also some variation in structure and function. For example, male brains in this sample were slightly larger by average volume than female brains, but there was also greater variability within the sample of male brains relative to the sample of female brains. In some regions, female brains showed slightly higher levels of connectivity. Ritchie and colleagues are careful to point out that these are small differences. The effect on cognitive processing, if any, will be small and not likely to have an impact on what people can do for a living, how they work, and how they interact with each other. Male and female brains do differ, but they are, for the most part, extremely similar.

---

5 In this book, I'll use the general convention of *sex* referring to genetically determined biology and *gender* to refer to the human sexual identity, which is usually strongly correlated with genetic sex, but not entirely.

Brains differ in size between people and they differ in terms of their internal make up. But brain size doesn't tell us very much about thinking and behaviour. We need to talk a little bit more about what your brain is made of, about the different areas of the brain, and how these different areas are specialised.

White and grey matter

Broadly speaking the brain is made up of white matter and grey matter and these vary in density. Both are very important for cognitive function and behaviours. Grey matter tends to be what we refer to more often when we use expressions like 'he's got a lot of the old grey matter' and it is what we see if we're looking at a brain. Grey matter is made up of the cell bodies of neurons; the interconnected cells that we often think of as being 'wired' when we use metaphors like 'that behaviour is hardwired'. The metaphor that thinking is electrical activity is from the twentieth century and reflects the science and technology of the time. The average human cortex is made up of about 16 billion neurons, each with multiple connections to other neurons. In addition, there are billions more in the other structures of the brain and the connections between the brain and the spinal cord and the rest of the body.

Neurons vary incredibly in terms of size and structure. The area of the brain that is heavily involved with higher order thinking (the frontal cortex, described in much more detail later) is made up of cells that may only be a few microns or millimetres long and are densely connected with other neurons. Motor neurons, which are the cells that connect the areas of the brain that coordinate movement with muscles in other areas of the body, can be very long; up to a metre in length. Of course, much of that length is in long fibres that connect the cell to muscles. When I was about six or seven years old, this fact terrified me for some reason. Trying to visualise a single cell in my body that was almost as tall as I was resulted in a nightmare vision of something that looked like a giant squid living inside my body. Not the best image to have when you are trying to sleep!

White matter is fatty tissue and for the most part it corresponds to the connective tissue and the myelin that surrounds some of the neuron. Myelin is important for learning and surrounds the parts of the neurons that form the connections among other neurons (the axons and the dendrites). Myelin, being composed of tissue, helps to insulate the connection from other neurons. This insulation is important for connection speed. Generally, the more insulation the neuron has, the faster the electrical impulse can travel from one end of the neuron to the other and this will increase the general speed of cognitive processing.

The basic substance, the white and grey matter, is what you see when you look at a brain. The white and grey matter is created from a dense and thick mass of interconnected neurons. This doesn't tell us too much about thinking yet. What matters a bit more is how those neurons are grouped together and how these localised groups work. The groupings are usually referred to as lobes and there are four distinct lobes in the human cortex.

The four lobes

Your brain is not just one big mass. The brain's physiology and function are organised systematically into four distinct areas known as lobes. You probably already know a bit about these brain areas, but let's discuss each of them in some detail. Knowing something about the basic functional architecture of the brain will end up being very helpful in understanding the organisation of the mind and of thinking and behaviour. And before we go on, I want to say something about the tendency to use certain metaphors when we talk about the brain.

When we think about the brain, it's nearly impossible to do so without using metaphors. Probably the most common is the 'brain as computer' metaphor. In this metaphor, the brain itself is seen as the hardware or the machine whereas cognitive operations are more like software. What we call the mind is the result of running this cognitive software on the brain. This does not mean that the brain is literally running software, but it's a metaphor that describes the relationship between brain functions and structure. The computer is a common metaphor that initially became popular in the 1960s and is still used today. But there are other, older metaphors.

Another metaphor for brain and mind that I really like is what I call the *hydraulic metaphor*. This probably goes back at least to Descartes, who advocated a model of brain function whereby basic neural functions were governed by a series of tubes carrying 'spirits' or vital fluids. You might laugh at the idea of brain tubes, but this idea seems quite reasonable as a theory from an era when bodily fluids were one of the most obvious indicators of health, sickness and simply being alive: blood, discharge, urine, pus,

Basic Brain Anatomy

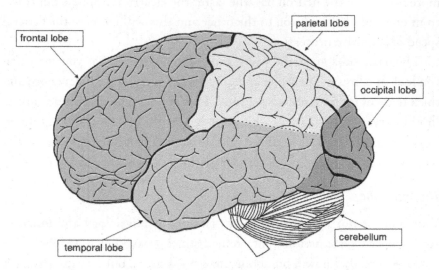

Figure 3.1 A side-view diagram of the human cerebral cortex, showing the frontal lobe, the temporal lobe, the parietal lobe and the occipital lobe. The cerebellum and brain stem peek out from the bottom.

bile and other fluids are all indicators of things either working well or not working well and when they stop, we stop. And in Descartes' time, these were the primary ways to understand the human body. So, in the absence of other information about how thoughts and cognition occur it makes sense that early philosophers and psychologists would make an initial guess that thoughts in the brain are also a function of fluids.

This idea lives on in our language in the conceptual metaphors[6] we use when discussing thought. Conceptual metaphor theory is a broad theory of language and thinking from the linguist George Lakoff. One of the basic ideas is that we think about things and organise the world into concepts in ways that correspond to how we talk about them. It's not just that language directs thought, rather, language and thoughts are linked. Our language provides a window into how we think about things. Chapter 10 will discuss this theory in much greater detail. We use metaphors, including this *hydraulic metaphor*, when we talk about the brain and mind. As a result, we

---

6 I will have a lot more to say about conceptual metaphors later in the book when I discuss language and concepts. Most of these examples correspond to metaphors in English, usually American English. Other languages will have other metaphors that may or may not overlap.

often talk about cognition and thinking as information 'flowing' in the same way that fluid might flow. We have common expressions like the 'stream of consciousness' or 'waves of anxiety', 'deep thinking', 'shallow thinking', ideas that 'come to the surface', and memories that come 'flooding back' when you encounter an old friend. These all have their roots ('roots' is another conceptual metaphor of a different kind!) in the older idea that thinking and brain functions are controlled by the flow of vital fluids through the tubes in the brain. And as you will see below, I'm going to discuss neural activation as a 'flow of information'. I might write that information 'flows downstream' or that there is a 'cascade' of neural activity. Of course, I don't really mean that neural activation and cognition are flowing like water, but like so many metaphors in our language, it's just impossible to describe things without using these expressions and in doing so, activating the common, conceptual metaphor that thinking is a fluid process.

For this tour of the brain, let's begin at a cognitive headwater, the place from which information from the outside world enters the cognitive stream. The *occipital lobe* is located at the back of your head, right up against the skull (see *Figure 3.1*). The primary function of the occipital lobe is vision, and since humans are primarily visual creatures, the occipital lobe really does seem like the headwater for the flow of information. That may seem counterintuitive at first: your eyes are at the front of your head and visual information is processed at the back of your brain, but as we'll see later in Chapter 4, the visual pathway leading from your eyes to the back of your brain actually helps to process information for you, so much so that by the time the visual information that was taken in by your eyes reaches the brain, it had already been partially processed into chunks with some basic information about location, colour and rough contour.

Information from the visual pathway enters the brain at the very back part of the occipital lobe and the neural information flows forwards in the brain. As this happens, a cascading network of neurons breaks down the information into perceptual features like edges, contours, spatial location of edges and contours, and movement. As the information flows further along in the occipital lobe, the information is processed further into more complex features like angles and conjunctions. These are known as 'visual primitives' and these eventually can be processed further downstream as letters, numbers, shapes and objects.

At some point, however, the streams diverge. Information from the visual system splits off and flows along a pathway to the top of the brain to the *parietal lobes* and another pathway leads to the *temporal lobes*. The parietal lobes are concerned with sensory and spatial integration, and for that reason the visual stream from the occipital to parietal lobes is often referred to as the 'where system'. The parietal lobe processes information from other senses too, including touch information from different parts of your body. There are sensory neurons that connect the parts of your body (lips, tongue, finger tips, abdomen, etc.) to areas in the parietal lobe and, not surprisingly, there is relatively more of the cortical region dedicated to processing tactile information from sensitive areas like your fingers and lips and relatively less from areas like your lower back. Tactile sensitivity is correlated with how much cortex is dedicated to that area.

Neurons that were activated by visual information in the occipital lobe also send information along a pathway to the temporal lobe. This is sometimes referred to as the 'what system' because this is where we learn to name and label things and to form concepts. The temporal lobes are located on the sides on your head, behind your ears. This is convenient because that's where auditory information is processed. This is also an area that is critical for memory. A subcortical (meaning below the cortex) structure called the hippocampus is involved in processing information in such a way that it can be activated again: aka a memory. Because the temporal lobe is concerned with audition and memory, it's also where much of our human language processing developed. As we'll see later, damage to this area can have profound effects on memory, on undertaking spoken language, and on the ability to recognise things.

At the front of the brain, behind your eyes, is the appropriately named *frontal lobe*. This area of the brain is responsible for motor activity such as moving your hands, lips and head. Within the frontal region, there is an area at the top your head called the motor strip that is right beside a section of the parietal lobe that processes sensory information. These two areas work together to send and receive sensory-motor information to and from every region of the body. There is also an area of the frontal lobe located right beside the temporal lobe that helps to generate spoken language. Language use involves several areas of the brain to coordinate the cognitive aspects of language with its spoken, visual and auditory aspects.

The very front of the frontal cortex is the area that I referred to earlier as the prefrontal cortex. This area of the brain is not unique to humans, but we are unique in how large, relatively speaking, our prefrontal cortex is. This area of the brain is responsible for planning, inhibiting behaviour and selecting what to pay attention to. The prefrontal cortex is also responsible for coordinating and regulating some of the functions of other areas of the brain. And as I described at the outset, this was the area that was damaged in the car accident that changed my friend's behaviour. The damage to her prefrontal cortex was enough to change her personality because it affected the way she planned, decided and selected behaviours. The other areas of her brain were fine, so she was able to speak, remember and perceive things. But she could not put all these things together in the same way that she used to.

Subcortical structures

I also want to discuss just briefly a few of the structures that reside under the cortex, below the part of the brain that you would see if you were looking at it. I'll be referring to some of these later in the chapter, and again when I discuss memory later in the book. So, we might as well describe them here.

The first of these is the *hippocampus*, which is located below the surface of the temporal region (see *Figure 3.2*). The hippocampus is found in many other species besides humans and it is responsible for creating and consolidating memories. The role of the hippocampus was discovered by the Canadian neuropsychologist Brenda Milner who is at the Montreal Neurological Institute. In 1953, she began to study a patient named Henry Molaison[7], who had been treated for his severe epileptic seizures by having portions of his temporal lobe removed. That's where the seizures had been originating from. The neurosurgeon, Henry Scoville, was known for pushing the edge a bit and even at that time the surgery was considered to be a bit radical. The surgery, known as a temporal lobectomy, did cure his epilepsy. But no one expected what was going to happen next. Henry was unable to form new memories. He developed one of the most severe

---

7 Henry was known only by his initials H. M. until his death in 2008 and was an inspiration for Guy Pearce's character in the film *Memento*.

cases of anterograde amnesia ever studied. This meant that although he remembered who he was, where he lived, and indeed all the intellectual and factual knowledge he had learned up to the time of the surgery, he was unable to commit any new things to memory. Each day was essentially the same day for Henry.

It was such a curious and perfect case. Henry was above average intelligence and retained all of his language ability and memories. He just lacked this one important ability. After the surgery, Henry was a subject and participant in many extensive cognitive neuropsychological studies with Brenda Milner as the lead investigator. Through careful experimentation, she was able to show that the hippocampus is the structure that allows the brain to create new memories. It is able to create the associations and connections needed to take an existing state of neural activation (what's happening now) and recode it so that it can be reactivated later.

As well, Milner discovered that not all memories need this hippocampal system. I'll discuss this in much greater detail later in the chapters on learning and memory, but she discovered that memories for new *actions* and *procedures* do not seem to require this system. Milner tested Henry's ability to improve on a cleverly designed test of motor memory, the mirror drawing test. In this test, a person has to trace a complicated design on a sheet of paper, but they can't see their own hands. Instead, they watch their hands through a mirror. This is not easy. If you have a mirror nearby, try to do this. Try to trace the outline of a shape or copy some words while watching your fingers in the mirror. Impossible!

But here's the thing, if you do this for a few minutes, then take a break, then try again with the same shape, you will be a little better. It will still be difficult, but you will perform better. And if you tried this for a few days in a row, you'd be even better still. Your sensorimotor system would form new connections and new associations and you rely on this sensorimotor memory to get better and improve. What Milner discovered was that Henry improved with time just as well as study subjects who did not have amnesia and who had fully operational hippocampi. Henry never had any recollection of having done this mirror drawing task before, even if he had done it earlier that day. The explicit memory of the event was not there, but he improved, which indicates that the perceptual and motor memories were there and were being used. It seems that the hippocampus is critically important for forming new memory traces for events and facts, but not for forming new memory traces for how to do things. This discovery, and many similar discoveries by Milner, have helped to shape how we understand memory today.

Subcortical Structures

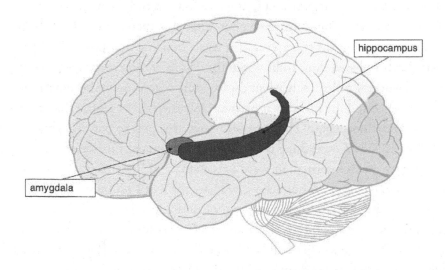

Figure 3.2: The location of the hippocampus and amygdala within the cortex.

The story of the discovery of the role of the hippocampus and its role in memory is one of the great discovery stories in cognitive psychology. The other subcortical structure I want to discuss, the *amygdala*, does not have such a story. Along with the hippocampus, the amygdala is part of a whole complex of subcortical structures sometimes known as the limbic system. This system includes several structures that are common to vertebrate brains and are quite similar across many mammals. The limbic system includes the hippocampus and amygdala along with the thalamus, hypothalamus, mammillary bodies and other structures. There is no consistent agreement about the structures that make up this system. Some neuroscientists avoid using this term. But these structures all seem to work together to accomplish the goals of learning, memory and, in the case of the amygdala, fear and emotional control.

So now that we know a bit about the structure of the brain, how information is sent, and how the different areas of the brain are somewhat specialised for different things, we can talk a bit more about how these areas work together. A good way to see how the brain systems work together in an individual is to look at case studies.

Case Studies

Have you ever walked past a building that was recently demolished and it occurs to you that you don't remember what was there? It can be difficult to remember what was there, even if it was a building that you walked past every day, because we don't think about the details of the streetscape when it's in good working order. Only when the building is gone do we become aware of its absence and the role that it played. Only then do we realise that it was part of something bigger. The same idea seems to apply to our thoughts and behaviours. When everything is working as it's supposed to, we really don't notice how much information we're processing and how many actions our brains are carrying out simultaneously. That's the way it's supposed to be. But when something is damaged or missing, we notice its absence as it helps us to see the bigger picture and the larger system. Case studies, patient studies and other similar designs help to explain how cognition works by looking at what goes wrong when a specific piece of the system is missing or broken.

Let's return to the story of my friend who was in the car accident. The force of the trauma that damaged her brain would have caused damage to the entire brain in the form of a massive concussion. This general damage might lead to general deficits, and indeed the initial outcome was a coma and a total loss of consciousness. Even after she came out of the coma, she was not fully present, and she lost access to language for a while. But language and communication ability were restored with time and some general healing. But the area of the brain that was most heavily damaged and the area that was partially removed was the prefrontal cortex. More specifically, she lost the most frontal part of the frontal lobe, including a section of cortex above her left eye (her left eye was also damaged). The area behind the eyes is known as the orbitofrontal cortex. This region is understood to play a role in decision-making, inhibition and people's ability to understand what behaviours are appropriate in social situations. Damage would result in changes in a person's ability in those areas. My friend lost a part of the brain that is responsible for many self-regulatory behaviours, for making decisions and for complex social interactions. Without the orbitofrontal cortex, she was still able to learn and remember, carry on conversations and still seemed intelligent, but she just acted differently. Without the orbitofrontal cortex, she seemed like a different person.

Case studies like this are fascinating because the damage and the sudden deviation from the way things are supposed to be highlights the link between brain structure and function. We have to be careful to interpret these effects within a larger context, though. Does behaving atypically mean that the orbitofrontal cortex is the region of the brain that controls your personality? Or does that mean that it is the only part of the brain that is responsible for planning, inhibition and decision-making? Not necessarily. What the pattern of damage and the resulting behavioural changes tell us is that the orbitofrontal cortex is part of a system that is involved in our ability to make decisions and inhibit behaviour. It also suggests that how we perceive a person's personality is made up in part by how we perceive a combination of many complex behaviours, including those abilities. If this combination changes, if this balance changes, we no longer perceive the same personality. This says just as much about the subtleties of personality and our perception of personality as it does about the complexities of the brain and behaviour.

The example of my high school friend is a good case study to begin thinking about how one area of the brain affects a set of behaviours. In this case study, the primary behaviours that were affected had to do with inhibition. You may know other people who have suffered damage to one area or another and who showed very different kinds of behaviour changes. For example, if someone suffered a stroke that damaged part of the temporal lobe, we would notice that their speech might be damaged as well. They might speak very slowly or have difficulty speaking in complete sentences. If there is damage to the pathway between the occipital lobe and the temporal lobe, the person might not be able to name objects that they are looking at. That is, they can 'see' the object, and they can grasp for it appropriately, and may even be able to name it if they pick it up or are holding it, but they just cannot come up with the name when they are looking at the object. Because there is a general localisation of function in the cortex, because there are specialised areas and pathways, damage to different areas will result in different behavioural changes. When studied more systematically, these case studies tell us about how the brain works and how brain structure is connected to function.

Phineas Gage

The most famous case study in cognitive neuroscience is probably the one of Phineas Gage, who was a rail company worker in the United States in the 1800s. His job was to help with demolition as the rail company cleared the right-of-way for their railroads. Gage worked with explosives. He would drill a hole in the rock, put explosives in the hole, and then tamp everything together with sand using a heavy metal bar called a tamping iron. The tamping iron was over a metre long and 3cm in diameter. It sounds dangerous even on paper. One day Gage was working to clear an area through some rock in Vermont and was tamping the explosives into the hole when the charge went off. The force of the explosion turned the tamping rod into a projectile. It shot back at Gage and hit him directly in the face.

The force of the impact drove the tamping rod into Gage's mouth, upwards behind his left eye, and right out of the top of his head. The rod blasted through Gage's head and flew an additional 8 metres with part of Gage's brain on it. Meanwhile Gage fell backwards and convulsed. But surprisingly, he could move and talk and was alert as he was taken to a hospital. The treatment would have been very different from modern medicine. For one thing, no one really understood how the brain actually worked to control behaviour. So, he suffered massive swelling and concussions and nearly died but after weeks, he stabilised, and the risk of death was over. After time, he was able to try to get back to life, but as you can imagine, it was not easy. Like my friend in the car accident, Gage had changed.

There is a famous quote attributed to Gage's physician at the time, Dr. Harlow. Harlow writes:

> *The equilibrium or balance, so to speak, between his intellectual faculties and animal propensities, seems to have been destroyed. He is fitful, irreverent, indulging at times in the grossest profanity (which was not previously his custom), manifesting but little deference for his fellows, impatient of restraint or advice when it conflicts with his desires, at times pertinaciously obstinate, yet*

*capricious and vacillating, devising many plans of future oper-
ations, which are no sooner arranged than they are abandoned
in turn for others appearing more feasible. A child in his intellectual capacity and manifestations, he has the animal passions
of a strong man. Previous to his injury, although untrained in the
schools, he possessed a well-balanced mind, and was looked
upon by those who knew him as a shrewd, smart businessman,
very energetic and persistent in executing all his plans of operation. In this regard his mind was radically changed, so decidedly
that his friends and acquaintances said he was no longer Gage.*

This account probably stretches the truth a bit, as nineteenth century medicine sometimes did. And it's also believed that the picture painted by Harlow of what Gage was like before the accident is not quite accurate. No one had really observed and detailed his behaviour before the accident. He was a railroad excavator, which is not quite the same as a 'shrewd, smart businessman'. Nonetheless, there was enough information from family that corroborated this general picture of a man that was fairly average before the accident but changed into a different person after, one more childlike and with less inhibition. This general picture is, on the surface, somewhat similar to the general changes I noticed in my high school friend that I discussed earlier: the change in the personality, increases in saying or doing inappropriate things and the difficulty planning. And furthermore, the lack of inhibition is comparable to what is observed now in patients with frontal lobe damage. And later, post-mortem reconstructions of the damage seem to confirm this. Gage's case history showed a direct connection between the brain and some, but not all, complex behaviours. The explosives may have inadvertently launched the tamping rod through his head, but Gage inadvertently launched the field of cognitive neuroscience.

Despite the shortcomings of contemporaneous neuropsychological accounts of Gage and the 'Ripley's Believe-it-or Not' quality of the Gage story, it is an important one for the history of neuropsychology and cognitive neuroscience. Prior to Gage, there was a general understanding that the brain is responsible for thinking, language and planning behaviours, but next to nothing was known about the organisation of the human cortex.

The Gage case made it clear that cognitive and behavioural functions were localised. The Gage case also showed, albeit in a crude way, how brain lesions can lead to functional dissociation.

In cognitive neuroscience, a dissociation happens when damage to an area of the brain results in the loss of some function but leaves other functions unimpaired. Gage's case shows what is known as a *single dissociation*, meaning that one area is damaged, and one function or cluster of similar functions is damaged but other functions are not. This tells us that there is some relationship between the damage and the function, but it does not entirely rule out other explanations for the functional changes. This single dissociation only tells us about one area and one function.

A *double dissociation* can be shown when two different patients are compared. For example, Patient A may have damage to the neural pathway that connects the visual cortex to the parietal lobe (see *Figure 3.1*). Patient B may have damage to another part of the visual pathway that connects the visual cortex to the temporal lobe (again, see *Figure 3.1*). Given what you know already about the visual cortex in the occipital lobe and the functions of the parietal lobe and temporal lobes, what pattern of behaviour would you expect each patient to exhibit? Patient A will likely be able to name objects that are presented but would have difficulty grasping for them. Patient B would be unable to name objects but would be able to grasp them correctly and even identify them by name when touching them. This is an example of a double dissociation because Patient A shows function changes in one behaviour but not the other, and Patient B shows exactly the opposite pattern. The double dissociation is usually considered to be fairly strong neuropsychological evidence because it helps to rule out alternative explanations for functional differences between two patients and it shows how a single area can affect one process and not the other.

David and the Capgras delusion

There are some cases in which a double dissociation can even be observed in a single patient. A double dissociation in cognitive neuroscience means that damage to one area affects one observable process but not another. At the same time, damage to another area might affect that second observable process and leave the first one unaffected. One of the most unusual neuropsychological cases is that of the Capgras delusion. Capgras, also known as the 'imposter delusion' is an exceedingly rare cluster of symptoms in which the patient recognises familiar people, a spouse or a parent for example, but does not believe that the person is who they say they are. That is, they admit that the person looks the same, they have no trouble identifying them, they have no visual deficits, and no damage to any visual areas. They just don't believe their eyes. They suffer from a delusion as a result, convincing themselves that someone else has taken the place of and is impersonating their loved one.

Imagine how frightening that must be for both parties! It must be truly terrifying to live with someone and then start to think that they were somehow replaced with an identical imposter. And for the person without the delusion, it must be frightening as well, knowing that you are now living with someone who believes you to be an imposter. Any attempt to convince them otherwise would probably only add to the suspicion.

For a long time, this delusion was thought to be psychiatric in nature, meaning that although it might be connected to some damage in the brain, the true nature of the delusion was in some deep problem in understanding. Or it might arise as a way to resolve Freudian conflicts in the relationship with the person. But it's now understood primarily as a specific set of cognitive and behavioural conflicts that come about as a result of very specific damage. This was discovered in part by a remarkable case study involving a young adult patient, David, and his neurologist Dr. Vilayanur Ramachandran (Hirstein & Ramachandran, 1997).

David lives in California and as a young adult, he suffered damage to the brain as a result of a car accident. Unlike the case of my high school friend, the damage was not localised to the orbitofrontal region but was spread throughout the brain. David was in a coma for five weeks. When he eventually regained consciousness, most of his abilities began to return and things like intelligence, language use and visual perception appeared to be unaffected by the accident and he regained those functions. As he was recovering, he lived with his parents. They noticed something unusual: he did not believe that his parents were actually his parents. He began to tell them that they were imposters. For example, he would tell his 'imposter' mother at dinner that the woman who cooked breakfast was his actual mother and was a better cook. When his father was driving him, he told him 'you are a better driver than my dad'. And this delusion was not just about his parents: he thought his house was an imposter house. His mother recounted a story in which David became upset that he was spending so much time at the 'imposter house' and that he wanted to go home. His mother could not convince him that it was his house, so she took him out the front door, they drove around, and then came back in the back door and David said, 'It's good to finally be back home'.

David and his parents were happy that the accident had not caused more extensive damage, but they were at a loss to explain this. They noticed that David often believed that the person he was with presently was the imposter while the memory of the person from earlier in the day was the actual parent. And he was not always under the impression that his parents were imposters. He knew that his parents still existed, but he was confused when he was interacting with them. He was not able to believe that they were the same people. One day, they noticed something incredible: David never seemed to have these delusions when he was talking on the phone with one of his parents.

At first, they thought that this might just be a coincidence, but it happened every time. The delusion only happened when David was looking at his parents, it never happened when he was only speaking to them. In order for the delusion to work, it required that David be interacting with his parents visually. This didn't solve everything, though. It was already established that David was not suffering from any visual impairments. He could recognise his parents in photographs, even though he didn't always believe that they *were* his parents in the photo. And, he had no difficulty recognising visual objects. The delusion was restricted to his belief that his parents or his house were not really his parents or his house.

Dr. Ramachandran came up with an ingenious way to test a hypothesis about what was damaged in David's brain and how it was affecting his behaviour and causing this delusion. David was presented with photographs of people, including his parents, while an instrument on his fingers measured something known as the galvanic skin response. Galvanic skin response, or GSR, is sensitive to emotional reactions. When you see a photograph of someone you know and love, minute changes in skin temperature and sweat change the chemistry of your skin. You cannot tell this is happening, but a sensitive instrument can. What happens in most people is that when they see a picture of someone they love, the

GSR detects a change relative to seeing pictures of unfamiliar people. What they wanted to know from David was whether his brain and body provided evidence that he *was* recognising his parents as his parents even if he claimed the opposite.

When David was shown these pictures, he didn't show much GSR difference between familiar and unfamiliar images. That is, it didn't seem as if his brain was able to react in the appropriate emotional way to the familiar face. It seemed as if David's brain was able to recognise his mother or a picture of his mother as a true representation of her from a factual, intellectual level, but his brain was not providing the appropriate emotional response. The pathway from the areas that are specialised to recognise faces (called the *fusiform face area*, or FFA) in the visual cortex to the temporal lobe was not damaged. As a result, David was able to recognise his parents and he had access to general information about who they were. However, the pathway that connected the face-recognition areas with the emotion centres in the amygdala (*Figure 3.2*) was damaged and so there was no emotional connection and no difference in emotion between familiar and unfamiliar faces. In other words, his brain recognised his mother but did not associate that recognition with the correct emotion. David had to deal with an uncomfortable reality. 'This looks like my mother, but I don't feel like that's my mother.' Because of that disconnect, David's cognitive system resolved the conflict by imagining a delusion. They also did the tests with just voice, and found that when David heard his parents, his brain reacted correctly, and they were able to detect the appropriate emotional response. There was no conflict, because the pathway from the auditory cortex was not damaged.

You might be wondering why he didn't just accept this change in emotional response. Why didn't he just accept that the emotional response was different? Why didn't his rational mind understand the nature of the disconnect just the way we are now? Why did his mind decide to override the conflict in his brain by creating a delusion? It turns out there was some additional damage that prevented him from resolving the conflict. There was minor damage to some of the prefrontal areas which affected some of his executive control and interfered with decision-making and planning. Without this rational ability to override the disconnect, his mind created the delusion instead.

The Capgras delusion is rare and it's still not entirely understood. But it does help us to understand how different areas of the brain work to carry out a task that most of us do automatically and without even thinking about it: recognising a loved one. The damage and subsequent behaviours are the result of the broken system, but we can use that pattern to fill in what we need to know about the entire system of person recognition. Thinking back to the earlier discussion on the flow of information, information from vision enters into the primary visual cortex and the FFA is activated when the visual information matches the configuration of a face. But all that tells you is that there's a face there. The FFA can activate areas in the temporal lobe so that you can recall the correct memories and names, and it also activates the emotional centres in the amygdala so that you can recognise the person at an emotional level. The executive control areas of the frontal (*Figure 3.1*) region help to coordinate the appropriate response to all of this information and to the person we just recognised.[8] I haven't said anything at all about integrating sound and smell. And some of my descriptions of the brain regions are simplified as well. In addition, there is always a lot more going on at any given time. You may be talking to five or more people you know in a group of ten and have to keep all these faces straight, assign a name, respond differently to each one, etc. The prefrontal regions allocate your processing capacity and awareness though attention, and I will have a lot more to say about that in the next chapters. When this system works, we never notice all the parts and pieces. But when the system is compromised or damaged, we can see more clearly how it all works.

Taking Pictures of the Brain in Action

Case studies like those above help us to understand how the brain works and how different areas of the brain contribute to behaviours by the logic of subtraction or dissociation. If a patient has sustained damage to one area, we observe what behaviours are affected, are no longer present or have changed. We can also observe what behaviours are not affected, are still present or have not been changed by this damage. And by examining the aggregate of several case studies or patient studies, we start to get a coherent picture of how the brain is involved in our thought patterns, actions and behaviours.

---

8 I am presenting a very simplified description of person recognition.

But trying to understand the brain and mind by studying how it changes when it's damaged does not give a complete picture and has several downsides. For one thing, brain damage, whether by blunt force trauma, tumour or stroke is not very precise. Often several areas are damaged. And any time there is enough impact to destroy one part of the brain, it can be accompanied by swelling and concussion damage to the entire area. For another thing, these are usually unique, isolated cases. A clear case study like David's Capgras tells us a great deal but because the case is unique, it can be hard to draw inferences to the entire population. Finally, in many cases we are lacking information about what the patient was like before the accident or stroke that caused the damage. This was seen very clearly with the case of Phineas Gage. We know that some of these changes in his personality were a result of his brain damage, but we don't really know what he was like before the accident. The few testimonials that exist are not reliable, and back in the 1800s, people were not usually tracking the cognitive and behavioural patterns of railway demolition workers. We aren't really doing that now, either, although we do have school records.

So even as case studies are helpful, we need other ways to understand the brain. And in the late twentieth century, it became possible to measure brain activity in normal, healthy participants while they think, perceive, respond and behave. This has had a profound impact on the field of cognitive neuroscience, the field of psychology in general, and even in popular media. We can now observe the brain in action.

There are a number of ways to measure or image the brain, but I want to focus on two broadly defined techniques. The first examines the electrical activity in the brain and is able to detect very quick and immediate changes in response to something. The second technique measures blood flow in the brain and is able to make fairly accurate measurements about location: which areas of the brain are relatively more or less active during a cognitive task or behaviour. I'll discuss each in more detail and then we'll discuss how these techniques have revolutionised the study of psychology and behaviour.

Measuring electrical activity

As I discussed earlier, the neurons in your brain connect with and communicate by way of electrochemical energy. The connection between neurons is chemical: neurotransmitters, but the neuron 'fires' by sending an electrical pulse called a potential. When a neuron receives input (from another neuron or a sensory cell) above a certain threshold, it will generate a brief charge of electricity that travels the length of the cell and causes neurotransmitters to be released from the other end that will propagate the activity to another neuron. It's no wonder that the 'computer metaphor' for our brain is so compelling: in both cases, we have small units that communicate via electricity.

Techniques for measuring this electrical activity have been around for a long time. Almost as far back as Gage's time, in the late 1800s, physiologists realised that electrodes could record this electrical activity. A German physiologist and psychiatrist named Hans Berger recorded the first human electroencephalogram (EEG) in 1924 but there were not many research or clinical needs for this technology until later into the 1960s. These recordings from the twentieth century were generally restricted to long recordings of whole brain activity and contributed to what we know about the brain's activity during sleep, dreaming (REM), alert activity, and agitated states. This tells us about the brain but very little about how the brain works to carry out cognitive tasks.

However, if the EEG measures are combined with a stimulus event, it can provide more information about how the brain reacts when you hear something, see something, or experience something. This measure is called an Event-Related Potential (ERP) because the electrical potential is connected to an event. This technique has been known since the middle of the twentieth century, but it wasn't until well into the 1980s and 1990s that the technique became more widespread, helped in part by availability of sufficient computing resources needed to carry out the analyses. As with many of the advances in cognitive psychology, the computer played a big role.

Here's how the technique works. The participant is seated in front of a computer display (for visual studies) and they wear a set of electrodes on their head. This looks a little like a swimming-cap with about twenty wires coming out of the top. The wires are connected on one end to electrodes that sit next to the scalp and these are connected to a computer interface for recording the electrical activity. The participant then does some perceptual or cognitive tasks while the electrodes record brain activity. Because the computer is coordinating both the presentation of the stimulus and the recording of the stimuli, it can record the electrical impulses for a specific area immediately after the person sees an image, makes a response, or reads a word. This tells us how the brain responds to something even before the person is consciously aware of it.

One of the most robust findings in ERP research involves showing how the brain reacts to seeing something unexpected when comprehending a sentence. For example, a person might be asked to read a simple sentence on the screen that ends in an expected or an unexpected way. An expected sentence would be 'The cat caught a mouse' and an unexpected sentence would be 'The cat caught a mountain'. These are nearly the same, and the phrase 'the cat caught' creates an expectancy for 'mouse' which is either resolved or not resolved. When the person sees the unexpected sentence, there is a larger spike in negative voltage about a half a second after hearing the unexpected word. This is called the 'N400' component because it is negative voltage spike (N) about 400msec after the event (400). In this case, it seems to be related to the relationship between the words being used in the sentence and the concepts stored in memory, which is a topic I will discuss in more detail later in the book.

Both EEG and ERP have had an impact on research and clinical use. One that many parents might be familiar with is the Central Auditory Processing Disorder (CAPD) which is an umbrella term for symptoms related to listening and hearing, often in a school setting. Children who have been diagnosed with CAPD may have difficulty switching from what they are doing to pay attention to something being said to them. Since children may experience difficulty listening for several reasons, it can be a challenge to diagnose or understand this particular disorder, but ERP recording helps to show that the brain may not be reacting to auditory input as expected.

Another recent application of EEG is the development of wearable technology that can record EEG while a person is doing something else. A Canadian company called InterAxon invented a small headband called the Muse. The Muse looks a little like a headset, but it has sensors that run across your forehead and above your ears. The Muse is designed to help people learn how to meditate; it works by recording the EEGs from the frontal area of the brain and it connects to a smartphone via Bluetooth. You can wear headphones while you meditate and the app on your phone can play sounds, like waves or a river. The ingenious part comes when you engage in meditation. The device records a baseline of your activity before you begin. As you meditate, it monitors the electrical activity in the frontal and temporal areas of your brain. If your mind starts to wander, and the Muse senses the change, it adjusts in real time the intensity of the sounds you are listening to. If you are in a state of mindful awareness, the waves might be very light but if you begin to let your mind wander, the sound of the waves picks up a bit, allowing you to use that as a subtle cue to bring your attention back to the breath (or whatever your focus is). The Muse is using your own electrical brain activity to provide immediate feedback. I've used these in my own research lab and it's an amazing device and technology. It's not hard at all to imagine how real-time EEG could be used to control other devices as well (like lights, robots and apps).

Measuring blood flow in the brain

One of the shortcomings of EEG/ERP as a research methodology is that it is not very precise with respect to location. ERPs can be recorded from regions on the scalp but cannot provide much information about structures in the brain or activation below the surface. EEG/ERP has very good temporal resolution but only moderate spatial resolution. However, techniques that measure blood flow in the brain are able to be far more precise. The most common methodology is fMRI which stands for functional (f) magnetic resonance imaging (MRI).

Neurons do not store energy so when they fire, they need to replenish glucose and oxygen and it is the job of the circulatory system to bring a steady supply. Oxygen-rich blood flows in, and oxygen-depleted blood flows out. In the early 1990s, a scientist named Seiji Ogawa discovered that oxygenated blood and deoxygenated blood have slightly different magnetic properties (Ogawa, Lee, Kay & Tank, 1990). This difference can be measured with a powerful electromagnet. The measure is referred to as the BOLD signal (Blood Oxygen Level Dependent). Areas of the brain that are relatively more active during a task, and thus require more oxygen, will have a different BOLD signal than other areas which are relatively less active. A participant in an fMRI study will be asked to lie inside a large electromagnet while they either see images presented or carry out other tasks. The magnet measures the BOLD signals in several areas, and these are analysed later to determine which areas of the brain were most active during that task.

As I pointed out earlier, your entire brain is always active, and the same is true during an fMRI study. In addition to thinking about the cognitive task of interest, you are also thinking about all sorts of other things: 'When is this experience going to be over? This is a huge magnet! Is this really safe? Where did I leave my phone? My back is hurting from laying here.' With all of this activity, how can the researcher isolate the BOLD signal for the task they are interested in? The most common way is a subtractive technique. The person is essentially scanned twice. For example, the first time, they could be scanned and not asked to think about anything in particular, and a second time they might be scanned and asked to imagine swinging a tennis racket. Both scans should be almost the same except that in one condition, the person is imagining tennis and in the other they are not: tennis imagery is the only difference. The next step depends on a powerful computer algorithm that subtracts the baseline condition from the tennis condition and the resulting scan should show which areas are relatively more active when the person is thinking about tennis. These should be, by the way, areas in the sensory motor area of the parietal and frontal cortex, the same areas that would be active if the person were actually playing tennis.

Research using fMRI, while still imperfect, has provided insights about the areas of the brain that process faces, songs, or are active when planning motor actions and when making complex decisions. Much of what is now known about the functional architecture of the brain (i.e., what happens where) has been discovered by fMRI or confirmed by fMRI.

One of my colleagues at Western University's Brain and Mind Institute, Dr. Adrian Owen, has pioneered the use of fMRI as a way to measure consciousness in patients in a vegetative state and even to communicate with them (Owen & Coleman, 2008). Vegetative state is what many people refer to as being 'brain dead'. For these patients, who are in a kind of coma, there is no sign that they are aware of anything in their surroundings. They may appear awake, but do not respond to voices, auditory or visual stimuli of any kind. For a long time, these patients were thought to be lacking any cognitive functioning or consciousness. It was thought that the brain was only operating at a low level, just enough to keep the patient alive.

However, Dr. Owen designed a revolutionary technique, with the help of modern brain imaging techniques, to measure conscious processing in some of these patients. First, he asked some healthy volunteers to imagine playing tennis while he scanned their brains, and not surprisingly, they showed increased activity to the sensory motor area when they were scanned with an fMRI. He then asked them to imagine playing tennis when the answer to a question was 'yes'. The research team would then ask them simple questions like 'Did you grow up in London?' If they did, they'd imagine playing tennis. In this way, he could look at brain activity as a proxy for a 'yes/no' response. Owen then tried this technique on patients who were in a vegetative state, thought to be brain dead and to lack consciousness. Amazingly, some of these patients, though not all, were able to imagine tennis and their brains reacted as if they were imagining tennis. They were able to use tennis imagery as a 'yes' response to personal questions and to answer questions about their surroundings. Many of these patients were consciously aware but could not respond or communicate. Dr. Owen's work has clear and profound implications for the care of these patients. As this technique is refined and made more portable (and adapted to other measurements techniques like EEG) it will give clinicians, care providers and family a way to communicate with patients and loved ones.

Conclusions

As humans, we are part of a larger system. We offload information to notebooks, our phones and the internet in order to remember things. We rely on other people to help us make decisions and solve problems. And a lot of our behaviours are guided by reactions to things in the outside world. But the brain is where everything comes together. The electrochemical activity in your brain defines who you are, what you're thinking about, and helps you plan behaviours. Until very recently, scientists knew very little about how the brain carried out these functions. But the development of cognitive neuroscience in the development of techniques that allow scientists to measure activity in the brain have given us incredible insight.

This field is rapidly developing, and within a few years of the publication of this book, some of the information may even be out of date. But the basic information about functional specialisation of different areas is likely to hold up. As you read subsequent chapters in this book, it's always worth thinking about how the brain accomplishes the more complex behaviours.

CHAPTER 4

Can You Trust Your Senses?

We exist in the world by way of our sensory systems: vision, touch, audition, etc. These senses tell us what we need to know about the world and are how we record what is happening now and what has just happened. These senses provide all the information we have about what is in front of us. They let us read, communicate and react. A good deal of the processing of sensory information into words, concepts, thoughts, memories and recognised objects happens further down the processing stream in the cortex, but the input comes directly from our senses. What are these sensory systems? How many are there and how do they work together? How do they work with internal representations and states, such as memory and thinking? Before I get too carried away with asking all these questions, let's try to answer at least one of them. Or rather, let's look at a common understanding for one of these questions and then explore if it's accurate.

I have a memory from my childhood in which I learned about 'the five senses' in my second year at primary school. As I'll explain later when I write about memory, this memory is probably inaccurate at best and possibly even completely fictitious. But I still have the memory. According to my school memory, these five senses are seeing, hearing, taste, smell and touch. That was it. You don't usually learn about the nuances of haptic perception, or the different ways in which vision is processed. We didn't learn much about sensory deficits or how individuals with loss of hearing or loss of vision process things. We didn't learn about sensory integration and how vision and hearing are combined. Just the five senses. This was an elementary school, of course, so you would expect the description to be a simple one. But most of us still think in term of five different and separate sensory systems, and we tend to think about sensation and perception with the same sophistication that we might have used in school. That is, we're not too sophisticated, and for the most part, we don't need to have a sophisticated understanding of perception. We trust our senses.

This early memory has some degree of specificity to it. For example, I remember that the classroom was on the left side of the hall (no doubt the result of a visual-spatial memory and visual perception). I remember that it was my first year at the school, having transferred from another school. I don't remember the teacher's name, but I can sort of remember some things about how the room looked, how it felt to be there. These are sensory memories. Now as I said, this memory is probably not true, despite how specific it is. There probably was not one single day when this was all laid out to me. Maybe it was in the first year, which seems more realistic, or even pre-school. Or maybe it was on television on *Sesame Street*. Maybe it was all of the above and I just combined them into the specific memory that I now have (and as we'll see later in the book, that's how memory works, by consolidating and reconstructing experiences). The point is that I don't really trust my memory about this. I can experience the event via memory, but I do not trust it to be very accurate. I have a current and real recollective experience that I do not really trust. I may not trust my memory, but I do trust my senses. You probably trust your senses too. But should you?

The phrase itself is a common idiom: 'you have to trust your senses', 'trust your own eyes', 'believe what you see', 'seeing is believing', 'I'll believe it when I see it', 'pics or it didn't happen'. That there are so many common expressions like this suggests that there is an underlying idea, or a conceptual metaphor which holds that 'perception is reality' or maybe 'perception is belief'. This idea is prevalent in our culture and so it comes through in our language. In fact, there is something ominous in being told that you should not or cannot believe what you see. George Orwell wrote, in his book *1984*:

> The party told you to reject the evidence of your eyes and ears. It was their final, most essential command.

When Orwell wrote this, it was supposed to be a nightmare scenario, a ruling party that explicitly tells people that what they see with their own eyes should not be believed. Beginning around 2015 and into the 2020s, well after the events of that novel were set, the modern idea of 'Fake News' has been propagated by many different factions and in several countries. A clear and recent example is the size of the crowd at the 2017 inauguration for U.S. President Donald Trump. President Trump claimed that it was the largest inauguration crowd in history, but photos taken from the top of the Washington monument strongly suggest that the crowds were much larger for the inauguration of Barack Obama.[9] Of course, from the perspective of where the President was standing, the crowds must have seemed incredible. From the perspective of people on the ground, it could have seemed very large (if you were in the front) or sparse (if you were standing many blocks away). The disconnect resulted in the President's own press secretary disputing the photographic evidence in favour of the official account and also resulted in many people questioning which version of the facts (or 'alternative facts' as one White House staffer put it) should be accepted.

While this may not be the same as the fictional scenario that Orwell was writing about, it still introduced uncertainty into what should be a fairly clear and direct observation. Because of this uncertainty, we wonder what we should believe. Being told that you can't trust what your see or read seems to many people to be a frightening and unsettling experience.

But should you trust your senses? Is it true that 'seeing is believing'? I argue that in many cases, the reverse is really more accurate, that 'believing is seeing'. I want to start this chapter with some clear examples of how you should not trust your senses: visual illusions. Then I will explain how your sensory systems and brain work and how these illusions are just exaggerated examples of how we perceive and understand the world. We do not see the world that is directly in front of us, but what we see is a reconstruction of what is in front of us, blended with what we already know. Understanding how the system works and why it works the way it does, can help to reduce the feeling of uncertainty when confronted with the possibility that things are not always as they seem.

---

9 There is no official number, but estimates using photos and counts of people riding the Metro suggest around 500,000–600,000 people attended Trump's inauguration and around 1,800,000 people attended Barack Obama's inauguration in 2009. No credible source has shown the Trump inaugural to be larger than the Obama inaugural.

The Study of Illusions

One way to show that things are not always as they seem is to examine sensory and perceptual illusions. The world 'illusion' comes from Latin by way of Middle English and its root is 'to deceive'. We usually think of illusions as tricks or deceptions. An illusionist is a performer who tricks the audience into thinking they see something other than what is in front of them. In the same way, we often think of a sensory illusion as our sensory system's attempt to deceive us. Though it might be more apt to describe the illusion as a deception that is created by a disconnect between what is activating the sensory input and how the rest of the brain is interpreting the sensory input. The illusion is the result of resolving a conflict between sensation and knowledge in favour of knowledge. In that way, it's not really a deception, but the result of an unconscious and often involuntary decision made in favour of prior evidence. Illusions show us that our brain and mind are working to make assessments and predictions about what we see. Usually these predictions align with the sensory information in front of us and we don't notice anything. When they don't align, we experience an illusion.[10]

Illusions exist in different modalities. There are auditory illusions, in which you might perceive hearing something that did not happen. For example, your brain will fill in missing speech sounds to complete words. This is an illusion in the sense that you perceive what was not there, but it's also a helpful prediction and it actually keeps us from making errors in hearing what people are saying. There are touch illusions in which you might perceive the touch of something that is not there. For example, the 'phantom buzz' with your smartphone occurs when you think about receiving a notification and sense a buzz or vibration from the phone. These might all be examples of illusions, but the process for their creation is different. Some seem easier to dismiss than others. Let's look at a very simple visual illusion first, one that is clearly a conflict between what you are sensing and what you are perceiving and one that is very difficult to dismiss.

---

10 It's possible to argue that everything we see and hear is an illusion, in the sense that what we see and hear is a reconstruction based on our assumptions and concepts. But I will restrict the use of the term 'illusion' to mean cases in which the understanding of what we see is not the same as the objective sensory input.

The Müller-Lyer illusion

One of the most fundamental and robust illusions is the Müller-Lyer illusion. You have definitely seen the Müller-Lyer illusion even if you do not know the name. In the figure below (*Figure 4.1*), there are two horizontal lines, each flanked by a pair of arrowheads pointing out or by a pair of arrows pointing in. The line on the top, the one with the arrow pointing in, appears to be longer than the line on the bottom, the one with the arrowheads pointing out. I'm sure you have seen this already, and I'm sure that you know that both lines are exactly the same length. And if you are not sure about this, please take a piece of paper and measure one and compare to the other. They are exactly the same length, and yet they do not appear to be the same length. I have probably seen this figure or a similar figure hundreds of times and I know that the lines are the same size and yet despite my knowing, I cannot see them as the same size. That is the simple illusion. Both lines are the same. You know that both lines are the same. And yet both lines do not appear to be the same. The question is, why does one line appear to be longer than the other?

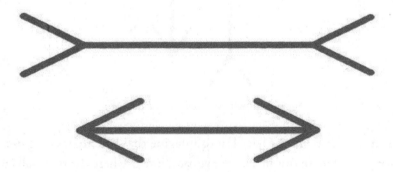

Figure 4.1 Müller-Lyer seen from a horizontal angle.

Let's look at how this illusion works in more detail. First, at the lowest possible level in the information processing stream, the lines will be the same. The image on the retina of your eye will be an accurate reflection of what is in the outside world. I will describe much more about the retina later in this chapter, but for now, let's agree that the retina, which is the back of your eye, is the structure that all the light and visual information first activate. The two lines will take up *exactly* the same space on the retina because they actually are the same. And the two lines will activate the primary visual cortex in the same way. In other words, from the 'bottom-up', these should be the same. So, where does the conflict come from? It comes from the 'top-down' knowledge and assumptions that you have about how the world works and how objects exist in three-dimensional space. This top-down influence overrides not only the sensory input coming from your eyes, but also your own, personal knowledge that these lines are the same. This top-down knowledge is deeply ingrained, and in some cases, it is innate, having been selected for by eons of evolution. These are deeply ingrained assumptions about the visual world.

The role of these deeply ingrained assumptions might be easier to see if we turn the image on its side (*Figure 4.2*).

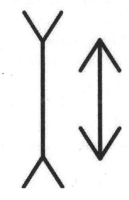

Figure 4.2 Müller-Lyer seen from a vertical angle.

Look at the figure on the left. Try to imagine that you are looking *into* the corner of a room. In this image, the vertical line is where the two walls meet to form the corner and the arrows might be where the walls and ceiling (or floor) meet. Can you see that? Now for the figure on the right, imagine that you are looking at the outside corner of a building. In this image, the vertical line is again where the two walls meet to form a corner, and the arrows are the top and bottom of the square building receding into the distance. In the figure on the left, the vertical line would be the farthest point from the viewer and in the figure on the right, the vertical line would the closest part of the object to the viewer. This would imply that the vertical line on the left, the inside of the corner, might be farther away that the vertical line on the right, the outside of a corner.

These ingrained assumptions about perspective and distance lead you to see one line as being closer than the other. This assumption activates two more deeply ingrained assumptions about the visual world. First, objects that are closer to you appear larger and take up more space on your retina. That's the way the physics of the eye work. Second, we understand that the same objects in the world tend to be of a constant size regardless of how close or far they are. We call that 'size constancy'. If you are looking at two people, one close and one far, the closer one will take up much more space on your retina. But you don't usually see it that way, what you see is two people the same size.

So, let's put all of these facts and assumptions together. You have two lines that will take up the same amount of space on the retina, but the context around them makes you think that one is closer than the other. Close things are usually larger on the retina. But you think you see a close line and a far line taking up the same amount of space in the retina. This could only happen if the far line were *actually larger* than the closer line. And so, you perceive the left line as being longer, even though you know it is not. The only way that you can simultaneously satisfy all the constraints is to create this deception. The deception which is really more of a correction. The deeply ingrained tacit knowledge you have about distance in three-dimensional space overrides the information coming directly from the retina and overrides your actual understanding. And so, you perceive the illusion.

In a way, the solution is not really an illusion at all. It's the resolution of a conflict between the information coming from your retina, and the way you interpret that information according to your assumptions and tacit knowledge about objects in a three-dimensional world. When there is a conflict, you almost always resolve it in favour of these assumptions. That's one of the problems with perception: it provides input, but that input doesn't make any sense unless it corresponds to something we already know about. What's more, we depend on our vision and object recognition systems to work as quickly as possible. We don't want conflict. We don't need conflict, because that would undermine the quick effectiveness of the visual system. In order to avoid a conflict, any potential disconnect will usually be resolved in favour of this deeply ingrained tacit knowledge. And importantly, this resolution cannot be overridden by belief, in this case the belief and knowledge that the lines are the same. The last thing we need is a visual system that can be willingly overridden by personal belief.

In short, this simple illusion demonstrates that we have deeply ingrained assumptions about the visual world. These deeply held assumptions cannot be overridden by belief or even by low-level conflict. They are usually correct, and they help to provide a consistent experience for us by smoothing out ambiguity and conflict that might come from within or without.

The Müller-Lyer illusion is an artificial illusion. It's an exaggerated attempt to create a conflict, and so it might not be too surprising that the visual system works the way it does. The natural world usually provides us with information that works with these assumptions, and not against them. After all, the assumptions come from the natural world. There are some examples of visual illusions that persist even in nature, however.

The Moon illusion

One of these is an illusion that you can experience on a clear night with a full Moon. It's called the 'Moon illusion'. If you've never heard of it, the Moon illusion is the perceptual experience of the full Moon rising or setting appearing to be much larger than the same full Moon overhead. When the Moon rises (or sets) it looks enormous on the horizon, but when straight

Moon Illusion

The moon appears larger at the horizon

Figure 4.3: An example of the moon illusion.

overhead it appears much smaller (see *Figure 4.3*). You would see the same thing with the Sun, but it's not advisable that you look at the Sun directly overhead. You would also see the same thing with a half or crescent Moon, but the illusion appears to be stronger when the Moon is full. How can this be? After all, the Moon is a set distance away from the earth, and whether or not it's on the horizon or directly overhead, it's exactly the same distance from you that night. This effect cannot be a function of distance and obviously the Moon does not change size. There must be something else, an interference between what your visual system is sensing and what you are interpreting according to your knowledge and concepts. This is probably one of the purest examples of not being able to trust your senses. There is no possible way for the Moon to change size, and yet it clearly appears to be larger on the horizon. How is this possible?

The Moon illusion has been known for a long time, and frankly has baffled people for many years too. Ptolemy, the Greek/Egyptian mathematician, was interested in the Moon illusion and recognised that it must be an illusion, perhaps caused by the differences in atmospheric refraction at the horizon versus the zenith overhead. Ptolemy also speculated on a slightly more bizarre explanation that when you tilt your head backwards, things seem further away. Later philosophers and astronomers offered alternative explanations. Cleomedes, a Greek astronomer, suggested that it could be due to our interpretation of apparent size. Arthur Schopenhauer, the nineteenth-century German philosopher, also suggested this hypothesis.

In the apparent size explanation, the Moon illusion is one that occurs because our tacit knowledge of objects in three dimensions conflicts with information from the sensory system. What are some of these tacit assumptions? One has to do with objects that recede towards a vanishing point at the horizon. We generally perceive and understand objects in the sky near the horizon to be further away than objects directly overhead. The horizon is the vanishing point, and so as things in the sky move farther away, they recede toward this vanishing point at the horizon. For example, if a bird flies overhead and continues in its trajectory towards the horizon, as it gets further away from you, it will get closer to the horizon line. It will also appear smaller as it flies further away. The same idea is true for objects on the ground, except this works in the opposite direction. Objects directly below your feet are close to you and if they get farther away, they recede towards the horizon line upward on the visual plane. Generally, things at the horizon are further from you than things directly overhead or directly below.

So how does this work with the Moon illusion? Let's break this down as we did with Müller-Lyer. We'll start with assumptions:

- Objects stay the same size regardless of how close or far away they are.

- Objects that are close will appear larger on the retina than objects farther away.

- Objects on the horizon are farther away than objects overhead and underfoot.

The Moon is, of course, very far away. We don't really have a sense of how far because we are not able to observe anything other than large celestial bodies from that great of a distance. Because it is so far, the Moon does not act in accordance with the receding horizon assumption. Unlike birds, airplanes or even clouds, the Moon at the horizon is the same distance as the Moon overhead. The Moon, unlike just about everything else you see, is not earthbound.[11] As a result, the earthbound horizon line assumption should not apply. But our tacit knowledge assumptions are only psychologically useful if they apply to all cases so that we can make fast assessments of the world. If we apply the assumptions to all cases, including the Moon, we still assume that all horizon objects are usually father away from us. This means we have to reconcile the fact that the Moon does *not* appear smaller at the horizon. That is, our tacit knowledge assumptions tell us that it is farther away at the horizon than when overhead, but it still occupies the same size on the retina.

One way to resolve this conflict is to assume that the horizon Moon must be both farther away *and* larger than the overhead Moon. If the Moon at the horizon were actually larger than the overhead Moon, both could cast the same retinal image. If the Moon at the horizon were actually larger,

---

11 The Moon is, of course, literally earthbound in the sense that its existence is bound to Earth's gravity. But it's not *on* the Earth and not part of the Earth.

it would resolve the conflict. Our mind prefers to resolve the conflict. And so, we perceive this horizon Moon to be larger. A larger Moon that is father away. This is just like Müller-Lyer in that our eyes see two things that are the same, but our minds are imposing some assumptions on these objects, namely that one must be farther than the other and therefore inexplicably must be larger.

This, of course, makes no sense. We know the Moon is the same size. Our eyes are not fooling us either, as they are taking in the same information. It's our mind that's confusing things. Despite the knowledge, the effect is nonetheless overwhelming. The next time you have a chance to observe a moonrise, notice how large it seems. Notice how it looms over the horizon. Notice that if you take a photo, it never seems to be as large. Then go back out and look again at the Moon when it is overhead and notice that it seems smaller and brighter. Notice, too, that we associate bright things with small (tiny points of light) and dimmer things with larger (big dark sea). Notice that we also associate smaller things with being up (small instruments make high notes, small birds, things that fly) and bigger things with things that are lower (large instruments make lower notes, large animals, things that can't fly).

These illusions suggest that the question posed by the chapter title, 'Can you trust your senses?' is not an easy one to answer. You might be tempted to say 'no' on the basis of these illusions, but these illusions arise because we do trust our senses. We have no say in the matter. Our minds have already decided. Our minds need to process sensory information at lightning speed so that we can make decisions about the world, select actions, carry out behaviours, and engage with the world. We can only do this if we impose some assumptions on what we sense. We trust that our senses align with the assumptions we impose. When they do not, we choose to trust our minds and not the conflicting sensory information. We choose to trust our minds because our minds already trust our senses.

In the next section, let's take a closer look at how sensory information is converted into perception. I want you to keep these illusions in mind, along with the idea that we need to have a cognitive system that is willing to make the occasional error in order get it right most of the time. That's a theme that's going to run through the book and that will re-emerge in many places.

The Visual System

Your visual system is a truly remarkable biological computer, built by evolution and natural selection (and built many other times in other species). When I first learned about the mammalian visual system, I was astounded at just how mechanical it is. Each part of the system carries out a small computation, processes one piece of information, and passes that information on to the next part of the system, like a computer system. Although it can be a cliché to say that the mind is like a computer, in the case of the visual system, it is accurate. As we will see, a great deal of the information processing takes place outside the brain. Note that when I say, 'information processing', I don't necessarily mean 'thinking', though the example could be described as 'cognition'. With vision, the extent of extracranial cognition is striking and due in part to how vision evolved. It's fairly accurate to say that the parts of the visual system that lie outside your cerebral cortex, your eyes and optic nerves, are a highly evolved cognitive system in their own right.

From the eye to the brain

As an example, I want you to consider the computational trajectory for the role of vision in the act of reading this book. I'll begin by discussing how light reflects off the page, is sensed by your eyes, and is perceived and processed by the visual system in the brain. We'll start with that example because that's how I assume most of the content in this book is being consumed and processed. I'm assuming that most of you are reading this in print, or maybe on a screen. To some extent, that's an assumption that is also a generalisation. It's what I assume to be generally true based on my experience with the world. But I also realise that my assumption does not capture everyone's experience. If you are listening to an audiobook of this, for example, then consider your experience of reading another book. If you are visually impaired, then my assumption definitely does not capture your experience. I am going to be discussing a case later about how the same brain areas involved in vision take over and make use of acoustic information in similar ways. For now, though, I'll assume that the majority of people interacting with this book are reading it visually.

Back to our example. The words on the page require some kind of light to be sensed and perceived. Words on a screen do as well, though the light is emitted by the screen itself. The visual system evolved to process light. Light is the only input. Many animals use light to get food, because they can't use light to make food the way plants can.[12] Many animals evolved to be sensitive to sunlight and we can use the same evolved systems to be sensitive to other light. Light travels nearly instantaneously, so it is a reliable source of information about what is around us. As a signal, there is essentially no time loss. Light bounces and reflects off things in the world and depending on the chemical properties of those objects, varying amounts of light energy are reflected and absorbed. That reflected light goes everywhere, but if you want to use it to navigate or plan behaviours, you need some mechanism to sense the light and convert it to information that you can process. That's what your eye does. Light from the Sun shines down on Earth, hits your page, and some of that light is reflected back in different ways (the ink does not reflect as much), and eventually that reflected light hits your eyes. All the ideas that have been known, all the stories that have been told, and all the court cases that have been decided can be experienced by anyone as a result of the same process: by light reflecting on a page or light emanating from a screen. The same light that emanates from a sun millions of miles away or powered by the electricity generated by the burning of fossil fuel (which is a by-product of long-ago sunlight) also carries the ideas and thoughts of people from long ago back to our minds in the present. Light, and our eyes, make this transmission of information and ideas possible.

Take a moment to acknowledge just how remarkable our eyes are. And keep in mind that the human eye is remarkable but hardly unique. Mammalian eyes have evolved in most species and are fairly similar to our eyes. Other species have evolved highly developed eyes that are quite different, and in many cases vastly superior to our own eyes in terms of sensitivity and acuity. For example, birds of prey, like the bald eagle, have extremely accurate vision. They can use their visual system to spot small prey from several kilometres away. Cephalopods (squid, octopuses and cuttlefish) have large, sensitive eyes that enable them to perceive details of their environment in low light conditions in the deep sea. Many insects have developed compound eyes. Some spiders, like the 'net casting spider', have night vision eyes so powerful that they look like little binocular goggles. These spiders have eyes so sensitive that they could in theory see galaxies

---

12 There's probably a deeper metaphor here, about how sunlight is being used by plants and animals to secure food, but I'm not sure what it is. And besides, many animals use other senses to get food, such as smell, touch and hearing.

that are too faint for our eyes to detect. Of course, this is a bit of a fanciful stretch, the net cast spider uses its eyes to detect prey and to guide the net that it casts. They do not eat galaxies and have no behavioural response nor concept for galaxies. They don't really see galaxies, despite the technicality of their eyes being capable of such a feat.

Let's return to our example of reading words from the page. The light reflected from the page of the book strikes your remarkable eyes and what happens next is even more remarkable. After passing through the outer membrane and cornea, the light enters the eye itself through an opening, the pupil. Your iris (the coloured part of your eye) surrounds the pupil and it can move to change the size of the pupil's opening (or aperture). This lets more or less light into your eye. When the light is brighter your pupil contracts to make a small aperture and when the light is dim, your pupil opens wide to let in more light. As the light passes through the pupil, it passes through the lens, which functions much like the lens of a camera. It's hard tissue relative to the rest of the eye, but it's still flexible and it is attached to muscles that can stretch it to change its ability to focus the light that is coming in. The lens focuses the incoming light through a clear liquid in the eyeball and into an image on the back of the eye (see *Figure 4.4* for details).

Structure of the eye

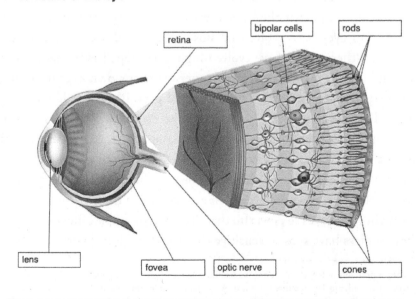

Figure 4.4: A cross section of a human eye with a closeup of the retina, showing rods and cones.

This image, which captures light from natural and artificial sources and primarily light that has been reflected from objects in the world, is focused onto the sensitive portion of the eye: the retina. And, as with a camera lens, the image will be inverted. The retina contains millions and millions of photoreceptors that sense and absorb the light and convert it into the electrochemical energy that neurons use. When you look into someone's eyes, their pupils appear to be dark, usually uniform solid black. This is because all of the light that enters the eyes is absorbed by the cells in the retina. When you gaze into someone's eyes, an image of your face, what we'll refer to as a 'retinal image', will fall onto the retina and that image is absorbed into the cells of the other person's eye. This image of your face is, in a very real way, now an indelible part of the other person's physiology. More than just a 'window to the soul', a person's eyes absorb images of everything that enters them.

Back to our retinal image. The image is focused on the retina of your eye, which is essentially that curved inner side of the back of your eyeball. It's covered by many millions of photoreceptors and each of them sit there, waiting to detect light. Human retinas have two kinds of photoreceptors: rods and cones. These are so named for the general shape of each kind of cell. The rods are rod-shaped. The cones are, predictably, cone-shaped. And as we'll see later, there are several kinds of cones as well. Although the entire image from the visual world in front of you falls on the retina, the lens focuses the image onto one special area of high acuity in the centre of the back your eye, called the fovea. As you are reading this book, the light and dark patterns of light that are caused by the reflection of light from the letters on the page, are then focused on the fovea. In other words, the thing you are looking at (a face, words, your coffee cup) casts its image on the fovea. Now of course, remember that your eye has no idea what you are looking at yet. Its ability to focus on an image of interest is governed both by properties of the world and by your knowledge and understanding of the world. I'll discuss more about this top-down processing later in the chapter and in other parts of the book.

Each of these kinds of photoreceptors has a specific job to do. The rods, which are located in greater proportion in the areas of the retina outside the fovea, are the more common of the two kinds of cell. There are about 90 million rods in each eye. They are also the more sensitive of the two kinds of cells. That is, they are sensitive to lower levels of light than the cones. There are two reasons for this high sensitivity, and one implication as a result of it. The first reason has to do with the chemical structure of the cell. These photoreceptors contain a chemical photopigment that is disrupted by light. In the rod cell, the photopigment is sensitive to a single light photon (a fact I just find to be astounding). When that happens, when light waves bump into this chemical, the rod can fire an impulse. It sends a signal down the line that 'light has been detected'. That's its only job. The second reason that the rod system is so sensitive is that they are connected to the next level of the system (the retinal bipolar cells) at a ratio of about 20:1. That means each bipolar cell has twenty rods out there, working to detect light for it. The end result is a very sensitive system with individual photoreceptors that can be sensitive to very low light levels, and a network where there are lots of rods working to detect light. But there are some implications as a result of this sensitivity. Rods are only sensitive to light, and don't have the capacity to respond to light of different wavelengths. Thus, they provide no information about colour. Another implication is that because they are connected to the next layer at a ratio of 20:1, that means that the next layer of the network has no information of where, within its field of twenty or so rods, the light was active. Now, that's still a pretty small region, but what it means is that what the rod system can gain in terms of sensitivity from the arrangement, it loses partially in acuity. If we want to be able to see colours, and to make out more detail, we need a different system and a different kind of photoreceptor. These are the cones.

Your eyes have fewer cone receptors relative to rods. There are only about 7 million cones in each eye. But they are nearly all located in the fovea, that area in the centre of your eyes where your pupil and lens focus an image. In addition, the cones are not as sensitive as the rods. They require more light in order to be activated than do the rod cells. But cones, though being fewer in number and less sensitive, contribute much to our visual experience. For one thing, the cone system is more accurate: it senses the world with great acuity. The greater acuity is a result of cones connecting to the next layer of the network, the bipolar cells, at a much lower ratio. Each bipolar cell connects to only about three cones. This means that, in contrast to the high sensitivity rods, the cones just don't pick up as much light and as a result, they are not as sensitive to low light situations. However, there is also a benefit to this low ratio. Because each bipolar cell has only a few cone cells providing input, there is more information about where the light source hit the retina. The network only has to figure out where in the field of three cone cells the light was detected. This is more accurate. Cones, relative to rods, have higher accuracy as a result of this. These cone cells are all located in the fovea, so you are focusing an external image on the cells that already have a higher acuity.

But these cone cells do more than provide a high-resolution image to the brain. There are actually three kinds of cone cells and each kind is maximally sensitive to light of a certain wavelength. Short cones are maximally sensitive to light of a shorter wavelength (light seen as blueish), medium cones are maximally sensitive to light of a slightly larger wavelength (what we see as green) and long cones are maximally sensitive to light of an even longer wavelength (light we see as red). The three kinds of cones, short, medium, and long, are sensitive to short, medium, and long wavelengths of light and thus enable us to detect objects that reflect back light at these different wavelengths. In other words, we use these rods to see colour.

The rods and cones don't know any of this. Rod cells don't know they are seeing only one colour. Short cones don't know they are seeing blue. Nothing in the retina really knows anything: the cells are just detecting light albeit in very specialised ways. This is a whole system. Each rod and each cone do very little. But the system constitutes our entire visual experience. It's the visual system that does the work. For example, the cones, though less sensitive and smaller in number, end up producing a rich visual experience as a function of the how the system is built. The network of cones to bipolar cells produces a different experience than the network of rods to bipolar cells. These two systems, each part of the visual system, evolved to achieve different goals. Rods detect more light and are more sensitive, so they detect smaller changes in the environment. Understanding these systems as systems and not just cells contributes to our understanding of the brain and mind in a larger sense. The power is in the network and the cognitive architecture. The power is in the computation level. The power is in the system, not the cells. This is true for all of the cognitive systems we will discuss, but it's particularly clear in vision.

Rods and cones are just the first step in this computation system. They receive the input from the outside world in the form of physical energy (light) and convert that to electrochemical energy for your brain to use and then send the signal to the next layer of neurons, which then send the information to the next layer of neurons and so on. The activation flows from your retina to the optic nerve (which is kind of like a big coaxial cable that pipes all the activation out of your eye). These neurons that carry the information from the back of the eye to the back of the cortex (your occipital lobe and the back of the head) partially cross over on the way. This partial crossover has an important function. At the level of the retina, you have two eyes (left and right) and two visual fields (the things to the left of you and the things to the right). But you see most of both fields with each eye. That is, your left eye sees things that are on your left but also sees things that are right in front of you, and also to your right.

To get this to happen, the visual information from each eye partially crosses over. In this crossover area, the optic nerve coming from each eye splits up so that the left visual field from each retina (from the left and the right eye) is combined. In other words, the information on the left side of the world is combined and the information from the right side of the visual field is combined. In terms of object recognition, it's more important to know which side of the world the object appears, rather than in which eyeball it is appearing. After all most things are viewed by both eyes.

Receptive fields

So far, we have followed the flow of visual information from the moment a photon strikes a receptor in your eye, to the optic nerve and the optic chiasm which is the area where the receptive fields cross over and join. At this point, the information from your eye is now ready to be processed by the brain. Your brain is not getting raw, visual information. The visual information has already been pretty heavily processed outside of the cortex by lower level visual systems. This processed information first activates the occipital area of your cortex, described in Chapter 3, which is the primary visual area at the back of your brain. The cells in your primary visual cortex are initially organised in what's called a *retinotopic map*. This means that the cells in the visual cortex are organised in a direct correspondence to the receptor cells in the retina. As a result, the neurons in the primary visual cortex can react to information as it is appearing in the retina. If you were able to record the activity of each individual neuron in the visual area, something that is possible in non-human animals, you would see a whole pattern of neural responses that have the same spatial organisation as the receptors, and by extension, the outside world. The brain would mirror the eye and therefore the outside world. At the lowest level, this processed information represents the outside world faithfully. It's direct, though not exact. That will soon change, as your tacit assumptions and knowledge begin to play a role. Your brain will not let this information from the sensory system go by without imposing some structure.

By the time the information from your retina and optic nerve reaches the primary visual cortex in the occipital lobe, the information is able to be coded for location and colour. The eye doesn't exactly know anything about colour yet, but your eye has rudimentary information about colour from the three different kinds of cones that respond to different wavelengths of light. Your brain decodes that information later and connects it to your concepts about objects and colours. Because the occipital lobe preserves the spatial arrangement in the retina, areas of activation that are adjacent in the world (because they are part of the same object) are also adjacent in the retina and therefore represented by adjacent neurons in the primary visual cortex. The eye does not know anything about objects yet, but because of correspondence of retinal to cortical cells, it is possible for the brain to make these assumptions later.

How do some of these assumptions work? How does your brain begin to take all the activation that's coming from your eye and begin to perceive features and objects? The answer starts with visual receptive fields. These fields begin to play a role outside of the cortex and are the primary way that your brain perceives features, shapes, letters and objects.

When I lecture about receptive fields at my university, I have found that it's always one of the first sticking points. For some reason, this seems to be the first major topic that often needs a bit more explaining. I'm not exactly sure why, but I think it's because up to now (in my class and in this book), we've been discussing simple statements of facts about the history of psychology, cognitive psychology in general, the brain, etc. This is the first topic that we hit that delves into the role of cognitive architecture. This is the first topic that deals with things at the computational and algorithmic level. To understand vision and cognition, we need to understand the problem of how to get raw visual information from the world to the eye and into the brain and how to turn it into objects. This means we need to understand the computations and algorithms that describe how visual information is processed. We need to understand how visual cells and neurons are built and connected to solve this problem. One way to solve the problem is with receptive fields. The arrangement of neurons and fields results in the extraction of features. Unlike earlier topics on the history of psychology or the basic anatomy of the brain, this topic is about cognitive processing and the computations that arise from connections. These connections result in the cognitive building blocks of perception and object recognition.

What is a receptive field exactly? These are cells that respond differentially to one pattern of visual activation and not others. Visual cells have preferences in these visual fields. These are like selective detectors. Think of this simple analogy, my car has several proximal sensors that detect if there are vehicles close to mine. Two of these are lateral proximity detectors that will detect the presence of an object in the blind spot that is on each side of my vehicle. If someone is driving up on my left side, for example, the sensor will activate an alarm to let me know. This sensor has one job. When there is an object beside me, it sends a signal. When there is nothing there it does not. This simple sensor has a receptive field that extends out from the vehicle. It detects things in that field and nothing else. It will activate to anything: cars, trucks or livestock. It does not care what colour or make the car is. It does not know anything, and it does not have to. On its own, this sensor with a receptive field does not do much. It just does its job. But if you connect that sensor to other parts of the system you can accomplish more. In a way, I am part of that system too. The side object detectors don't really detect other vehicles. I do that, by interpreting the output of those sensors. You could even build other sensors to detect other things in the environment and connect these sensors to detect more than one thing or a specific configuration of things.

A receptive field in the visual system is kind of the same idea as the vehicle detectors I described. At each stage of processing, from the ganglion cells that are behind the retina, through the different relays and into the primary visual cortex, neurons respond selectively to patterns of activation on the retina. For example, if you're looking at a white screen with a single vertical black line, the subjective experience you get is that you are seeing a single vertical black line, something like the number one.

Here is how it might look from the perspective of your brain, and bear in mind that I'm really simplifying. Each one of your retinal receptors responds to light and not to darkness. When it detects light, a receptor will fire in a certain way and when it doesn't detect as much light, it will fire in a different way. Now imagine a whole cluster of those cells, each one of them able to fire when they detect light and not when they don't. One kind of cluster, which we will call a 'centre on' cluster, fires more rapidly when light energy is activating the cells at the centre of the cluster, but not activating the cells at the periphery. That whole cluster fires more rapidly when the centre is 'on', but the surrounding cells are 'off'. If this cluster is connected to a single ganglion cell in the visual system, that ganglion cell is a called a 'centre on' cell, because its job is to detect light in the centre of its field (see *Figure 4.5*).

There are other cells which do exactly the opposite: they fire more rapidly as a cluster when the cluster of cells has more light energy striking the cells on the periphery but not on the centre. This cluster is connected

Receptive Fields

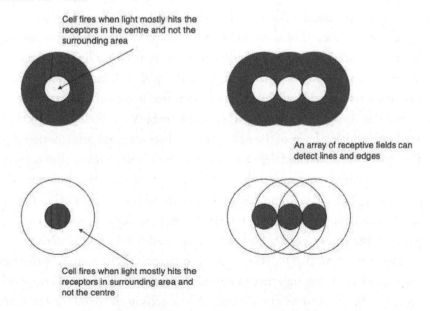

Cell fires when light mostly hits the receptors in the centre and not the surrounding area

An array of receptive fields can detect lines and edges

Cell fires when light mostly hits the receptors in surrounding area and not the centre

Figure 4.5: Receptive fields in the ganglion cells can be sensitive to light in the centre of their field or in the periphery of their field. This allows the cells to detect edges and lines that are light or dark.

to a ganglion cell and is called a 'centre off' cell because it fires when there is no light in the centre of its receptive field. If you were to take a network of these centre off cells, the kind that fire more rapidly when there's no light in the middle, more light in the periphery, and line them up, what would this array of centre off cells be able to detect? Well, you would get something that looks like an edge detector for dark lines. Each one of those centres is like a dark pixel surrounded by light pixels, and a whole array of those dark pixels surrounded by light pixels looks exactly like what you would expect. It's a dark line surrounded by white. If you connected this linear array of centre off cells to a single neuron, you would have a simple line detector. This line detector is built from clusters of cells, linked to other cells that are arranged in a particular way.

Your eye doesn't know that it's looking at a dark line on a white background, but it does have cells that specialise in sensing dark areas organised in a straight vertical line that is surrounded by white space. But you can see how powerful this architecture is. The system can make use of many, many different configurations of cells with different receptive fields. There are simple cells that correspond to light and dark bars of different orientation. These are like the ones I just described that respond to edges and lines. There are other more complex cells that correspond to light and dark bars of different orientation in different lengths, and there are other complex cells that correspond to light and dark bars of different orientations and different lengths that move across the visual field in a particular direction. At each stage, simple inputs are clustered together to feed into another layer of the networks to abstract more information. From the tightly clustered and dense network of rods and cones in the retina, you get lines, edges, contours, angles and movement. But think about what's happening here. You are imposing structure to the visual world but losing some detail. Each stage *abstracts* the information just a bit more. In order to perceive reality, we need to abstract things and to reconstruct things.

Let's try to put this together. We have described the way receptive fields work, how cells link together to connect to other neurons and create other receptive fields. And we also described how cells in the primary visual areas of the occipital cortex are organised in a retinotopic map. The next step in processing is the activation in your visual cortex of more specialised cells: cells that respond selectively to edges, lines, contours, and other visual features that are organised in your brain in a way that corresponds to how they are organised in your retina. Activation in the brain approximates the activation in the eye. Well before you know what you're actually looking at, your retina, your optic pathway, and your primary visual cortex have a fairly good representation of what information is in front of you. Your eye and your visual cortex have a detailed map of the pattern of light and dark information that you're looking at. And it really is like a map in a metaphorical sense, because the line, the edges, the colours, the gradient, are abstractions. They are a representation of what is in front of you, but not a perfect, exact copy. Some information is left out, some information is cleaned up and idealised.

At this stage in the flow of information, you don't really know what you are looking at. All you have is the detailed map that was created by the connections in your visual pathway. However, you have all the information you need to start pulling the features together into objects. This is where the story gets really interesting, and more complicated. Because in the cerebral cortex, you don't just have one visual pathway, you have two visual pathways.

Two pathways

According to the simplified account I laid out above, the initial stages of the visual system are data driven and computational. The arrangement of the cells provides your brain with information about where light and dark information is located in the visual field. Your brain is provided with information about colour, edges, conjunctions and motion. But you don't know anything about how to put those things together into objects yet. That's arguably the whole point of vision. We initially perceive the world as features, but we do not live in a world of features. We live in a world of objects. We need to be able to identify objects in the world. We need to be able to identify things so that we can navigate, interact and respond to things.

For example, if you're reading or listening to this book while you're at a café, you are probably sitting at a table with a coffee cup. Or if you are reading at home, you may also have a mug of coffee, a glass of water or cup of tea. It takes almost no effort to recognise the coffee mug or a teacup. But I want you to think for a minute about what object recognition is. To recognise the coffee mug, you need to segment those edges from other edges in the visual field and put them together. This begins with the data-driven, bottom-up feature detection that happens in the visual system, and that I just described. But that system will only get you so far. Recognition requires some knowledge about what coffee cups are. Object recognition also often requires a name for the object. When you see a coffee cup you recognise it as a member of the coffee cup concept. Your concept includes information about what it's used for, what its name is and what it's made of.

If you're sitting at a table or a chair and you have a cup of coffee or cup of tea, take a drink now. If you don't have a coffee mug, but you have a bottle of water, do the same thing. If you have neither, try to imagine yourself doing it. Regardless of which of these you chose, reaching and grasping for a coffee mug should have happened almost automatically. It shouldn't have required any conscious processing other than making the decision to reach for it. You probably did not have to think consciously about which direction your hands were headed. You probably did not have to think consciously about how to open up your hands just enough to fit around the coffee cup or how to close them around the mug, and you probably did not have to think consciously about how much pressure to apply to make sure that the cup did not slip out of your hands. Those motor actions were guided by vision. But they didn't necessarily make use of the object concept and name.

It seems that we recognise objects in two ways. First, we recognise things by name and identity. When we see something, and we can name it. In other words, we know what it is. But we also recognise objects by reacting to them and by behaving accordingly. It turns out there are two visual pathways that correspond to these two ways of recognising objects. These two visual streams collect the same visual input at the primary visual areas and split off in parallel in two directions. One of the streams, which is referred to as the dorsal stream or the 'how and where' stream, is a pathway that activates areas in the visual cortex through the motor cortex. This dorsal stream allows you to select the appropriate motor action in response to the visual environment. This could happen really quickly. It can also happen without consciousness. If someone throws something at you, you can put your hand up to deflect it without having to name the object. When you reach for that cup of coffee, the dorsal stream guides your hand and helps you grasp it in the appropriate way. In the same way that you can adjust the strength of your grip for a heavy mug of coffee or a delicate pastry. Different grip, different object.

The other visual stream is referred to as the ventral stream, or sometimes as the 'what system'. This ventral stream takes the activation from the primary visual cortex and sends it through to the temporal region of the cortex. This is where language is localised. And this is where you access words and link them to concepts. These two systems work together most of the time. Object recognition nearly always involves coordinating visual input with motor action and conceptual knowledge. These two systems talk to each other as well. When you think about the word 'tennis', activation from the language area of the brain can spread up to the dorsal stream area and activate some of the motor areas. And vice versa.

Neuroscientists have demonstrated that these two streams can also operate independently. For example, if someone has damage to the dorsal pathway, as a result of a stroke, they will likely have difficulty selecting the appropriate grasp for an object even if they can name the object. This pathway can be dissociated from the ventral pathway. If an individual has damage along the ventral pathway, they would be unable to name objects that are presented visually but they can usually select the right grasp for the object. This is a condition known as visual object agnosia. It just means that people can't name objects they see but they can behave towards them appropriately and in many cases, they can even name the object once they feel it. A person with visual object agnosia would be unable to name the coffee cup sitting in front of them as a coffee cup. But they would still know that it has coffee in it, they would still know how to reach forward, and once they pick it up, they would be able to generate the name coffee cup as a result of feedback from the way the object feels.

The visual system is complex and dynamic. It has been shaped by natural selection to allow us to interact with the environment. It is complex enough that even fairly substantial damage, such as a stroke, will not cause the entire system to collapse, but will just damage part of the system. In a stroke, the partial damage will result in the kind of general visual deficits described above, but other kinds of damage and changes to the basic system will result in even more interesting and fascinating changes. Some examples are very specific and yet each one can be explained by understanding the general cognitive architecture of the visual system. This architecture often operates even without primary input from the visual system.

Blindsight

For example, there's a specific case that one of my colleagues, Dr. Jody Culham, is working on. The individual, a woman from Glasgow named Milena Canning, is completely blind when she is looking at a stationary scene. She cannot make out any details, cannot recognise objects, cannot recognise people and cannot see letters or numbers. But she can see objects if they are moving. So, she has some brain damage. Based on what you already know about the flow of visual information in the brain, where do think the damage would be located? If you are thinking 'occipital' cortex, you're thinking correctly. She suffered a series of debilitating strokes that damaged her occipital cortex which interfered with her ability to see. In this patient's case, the entire occipital cortex was not damaged, but still enough of it to disrupt her vision. For all intents are purposes, she's blind.

Except she's not exactly blind. She had the sense that she could see fleeting bits of movement. Not exactly see, but she reported that she could 'sense' that she could almost see. When her doctors placed some chairs in a hallway, she was able to navigate along the corridor without bumping into anything. She could not see the chairs, exactly. She could not point to them or even report that they were there. But she altered her trajectory to avoid them. Milena's doctors in Glasgow thought she might have some very specific condition known as 'blindsight'. Blindsight is a condition in which a person who is otherwise completely blind can still sense and act on visual information. Usually it's the result of some residual undamaged cortex. But individuals with blindsight are not consciously aware of their ability. Milena, on the other hand, had a conscious experience of sensing something. Like a fleeting experience, a phantom of vision. But could there be other explanations? Maybe she's using sound or some other sense to guide her and not really vision. In order to determine what is going on, researchers needed to look at the brain. Her doctors in Scotland referred her to Dr. Culham at Western University in Canada.

Dr. Culham works at my university, in the same building and department. She's one of the world's foremost experts in understanding how vision guides actions. One of the things she wanted to do was test Milena's ability to perceive motion in controlled settings and to conduct a series of brain imaging studies to try to figure out what was damaged, and what was not damaged in the visual pathway. One test involved asking Milena to make motion discrimination judgements of objects that move on a screen. Milena was able to make accurate judgements about these figures, even though she could not make judgements about stationary objects. As researchers had initially expected, her vision was intact for perceiving motion. In fact, she reported being able to see an incredible number of moving objects:

> At the onset of the first block of a moving checkerboard, MC's eyes began watering and her cheek moved, as seen through the bore camera. At the end of that scan, when asked if she saw motion, she said, 'That was amazing. I could see thousands of things. I've never seen so much motion. I can't believe it! I was crying and laughing.' (Arcaro et al., 2019).

It seemed like under controlled settings Milena was able to see motion but not objects per se. When Dr. Culham and her team scanned the occipital cortex, they found very little activity overall, consistent with the idea that Milena was not experiencing much visual activity. However, Culham's team, using both functional and structural MRI, observed significant functioning of the middle temporal motion complex, which is an area of the visual cortex that specialises in some motion perception. Not only was this area spared, but it showed robust activation during some of the motion detection tasks that Milena was involved in. In other words, her subjective experience of being able to see things move (but not to see them in general) was supported by neuroimaging. She did not have a sense of vision per se, but her visual system could sense things moving and she could sense that as well. She was blind, but with sight.

Blindsight is rare, but not entirely unknown in people who lost their visual ability due to brain damage or stroke. As we saw in earlier chapters, the brain function for many abilities is distributed across more than one region, and damage to a part of that region will often result in damage to part of that ability. In this case, damage to Milena's visual cortex was extensive but not complete. As a result, her loss of function was extensive but not complete.

The brain and mind have a way of solving problems that arise as a result of damage or loss of input. In Milena's case, the input was still there but the area to process it was damaged. Let's look at an example where the visual cortex is fully functioning, but the input is not there. Or rather, the input is not as expected.

Echolocation

Many of the examples I have discussed so far deal with damage to some part of the system and what is lost as a result. But there are some fascinating cases that explore just how resilient the brain is and how function can be preserved. In this case, there is no brain damage. The visual cortex is just fine. But there is no visual input for this cortex. This is the fascinating story of blind echolocation.

Daniel Kish was born with a genetic condition that resulted in the total loss of vision at the retinal level. By the time he was a toddler, Daniel was completely blind. His visual cortex was not receiving any information from the optic nerve. There was nothing wrong with his visual cortex, but because it had no visual input, it did not have the visual input needed to construct a visual image or to carry out the basics of visual perception. But that's the amazing thing about people: they can adapt. Amazing thing about the brain: it can adapt, too.

Daniel soon began to navigate the world by sound. In particular, he spontaneously began to echolocate. He would make a sharp, repetitive clicking sound with his tongue and listen for the subtle changes in how the sound echoed back to be able to infer the description of different objects and obstacles in the world. Daniel has been doing this as long as he can remember but you can try it yourself. You won't be very good at it but try to see if you can tell the difference between two different rooms in your house. Walk into a large empty space, like a stairwell or a hallway, close your eyes, and try to make a sharp clicking sound (not a clucking sound). It should sound hollow and echoing. Now walk (eyes open or closed) into a smaller room, maybe one with a rug, carpet or some chairs. It should sound smaller, less echoing, muted. If you keep clicking and walk close to a wall, you will be able to hear the change. Of course, you already know what it looks like and so you're also filling in the details from your visual memory. But you should be able to hear the difference too. If you had a lifetime of practice and no visual images or memory to fill in the details, it should not be too difficult to imagine using a system like this to navigate.

As Daniel puts it in an interview 'I am sending out a signal and receiving the signal back and inferring things about the environment'. Which is not far off from the visual experience described earlier in the book in which light from an external source bounces off an object and sends the signal back to your eyes. In both examples, the signal is received and the person receiving it infers things about the environment. Daniel is just getting a different signal. He's getting an acoustic signal. Using this kind of echolocation, Daniel can carry on with his life as well or nearly as well as a sighted individual. He's able to cook, go for hikes, go shopping, and even ride a bike by echolocation. Bike riding is not very difficult for him, as long as he can hear the signals. With echolocation, he's getting a good picture in his mind about what is in the environment.

I used a visual metaphor just now, suggesting that Daniel has a 'picture in his mind'. But is it really a picture? Or is it something different? One possibility is that Daniel is using his hearing to navigate. This would mean that his navigation is purely sound-based. But another, more intriguing possibility is that he is using the visual areas of the brain, the same areas that are designed to recognise objects and to do visual navigation, but that they are just working on sound input, rather than visual input. How would you even be able to tell?

Some researchers at my university designed an interesting and creative experiment to find out. [13] Stephen Arnott, Lore Thaler and Jennifer Milne, along with Mel Goodale and Daniel himself, enlisted two other blind participants who had been echolocating for as long as they could remember (Arnott, Thaler, Milne, Kish & Goodale, 2013). They tested them extensively on their visual abilities and their echolocation abilities and they were in the same general range as Daniel. That is, no sight, no visual input to the cortex, and a highly-developed echolocation ability. Then they designed a study to find out what the brain was actually doing during the echolocation.

---

13 I won't always use research examples from my university, but I do so because I know it well. And in this case, it's such a great example. Mel Goodale, the lead researcher, is one of the world's foremost experts on visual cognition. There are few other places in the world where this work could have been done.

As we saw in Chapter 3, one of the most effective ways to measure brain activity is the fMRI scan. The fMRI measures cerebral blood flow to areas of the brain that are active during a cognitive task. By tracking the blood flow with magnets, you can tell if an area is implicated and active during a task. But there is one problem: fMRI is very loud and is carried out with the patient laying down, mostly immobile, with their head essentially inside a tube. You would not be able to do an echolocation task. Echolocators would not be able to hear very well over the noise and would not be able to echolocate anything anyway, beyond the inside of the tunnel. Now with visual cognition, the solution is pretty simple. You can show pictures of objects (on a screen) and record the areas that are being activated. The fMRI does not really get in the way. But how can you show pictures of an echolocation?

The researchers designed a creative way to do this. First, they asked the two research subjects to perform an echolocation identification of several different, easy to recognise objects in controlled settings. For example, they would have to identify large, smooth objects or objects with irregular edges that were covered in aluminium foil. These objects sound different because the surfaces of the objects reflect the sound back in different ways. That's the reason they look different too, by the way; they reflect back light in different ways. The object identification task was pretty simple, and the subjects were able to identify the objects reliably by echolocation. After the first echolocation task, they asked the subjects to do it again. So just as before, the subjects did what they always do to identify the objects. They made the sharp clicking sound with their tongue and listened for differences in how it sounded. But this time, while they were doing the echolocation, the researchers recorded the echolocation signals by placing very tiny microphones in the ears of each subject. Now when they made their clicking sounds, they sent their signal out, it bounced off the objects, and was reflected back and recorded by the microphones that were in exactly the same place as the subject's ears. In other words, the microphones picked up and recorded echolocation signals that would have the same sound that the subjects were hearing. In this way, there were essentially able to take an *auditory picture* of the objects. When they played these recordings back to the subjects that made them, the subjects were able to identify the objects as well as when they were doing an echolocation. That's because they were hearing their own clicking from exactly the same perspective as if they had made them live. It's really no different from seeing a picture that you took with your phone from your perspective.

Now that they had a recorded, auditory representation of the objects that was analogous to a photo, they could carry out an fMRI imaging study. They played the recording over noise cancelling headphones while taking an fMRI scan of the brain. What they found was remarkable. When the subject heard their recorded echolocation clicks, the auditory cortex was activated as expected. But so was the visual cortex! What's more, the visual areas showed evidence of the retinotopic map being activated in a way that corresponded to the shape of the objects. Later areas of visual processing were activated as well. For all intents and purposes, these subjects were seeing the world. Their internal, subjective experience was a *visual* one, despite having no input from their eyes.

This prompts a few questions about their subjective, cognitive experience and about vision in general. Do these individuals experience a visual image? If the neural circuitry that drives vision and visual imagery in sighted individuals is being activated in the same way in these blind individuals, does that mean that they *see* the objects in the same way that sighted individuals do? Possibly. The implication is there, but it's still very difficult to compare. One thing that was clear is that this effect seems to depend on how early a person's sight is lost, which seems to argue that this is not quite the same as vision. In blind individuals who learned to echolocate later in life (though still in childhood) the effects were less pronounced. It seems that if the visual cortex had been involved in vision, it was less able to take over for auditory object identification.

Another question that arises is what is the function of the visual cortex and pathways? This research suggests that the visual cortex is a general-purpose object identification cortex. It extracts object-bound features from a signal. It preserves some correspondence with the outside world, even as the information becomes more abstract. It tries to match input representations with existing patterns of activation (i.e. memory). It guides behaviour in accordance with these representations. It can direct cognition to names, concepts and memories. The visual cortex is one of the mind's primary links with the outside world. Its job is so important that if it does not receive visual input, it can learn to do its job with other information. Like sound.

Can You Trust Your Senses?

We have seen several examples of sensory input giving your brain an incomplete or even incorrect summary of the outside world. In some cases, like the visual illusions we discussed at the beginning, we might feel tricked because we know that what we see is not really what is there. In other cases, like blindsight or echolocation, it seems like the brain develops workarounds if there is not enough visual information to go on. What our brain actually experiences is an abstraction. A re-creation. A mix of the objective and the subjective experience.

We just don't see the world as it is. We see the world as a blend of what it is and what our brain needs to see. Should you trust your senses? Should you trust perception? Of course. Sure, occasional errors of perception or identification will happen. But these are infrequent and usually low cost. Our brains make these mistakes because perception depends on making assumptions, predictions and educated guesses about the world. These educated guesses are what the perceptual system is designed for. They help us to think and behave quickly and to perceive the world as we need to perceive it. Perception serves our behaviours, our goals and our drives. Perception keeps us alive. That's why we trust it. It's all we have.

CHAPTER 5

Attention: Why There's
Always a Cost

Everyone knows what attention is. It is the taking possession by the mind, in clear and vivid form, of one out of what seem several simultaneously possible objects or trains of thought. Focalization, concentration, of consciousness are of its essence. It implies withdrawal from some things in order to deal effectively with others and is a condition which has a real opposite in the confused, dazed, scatterbrained state which in French is called distraction, and Zerstreutheit in German. — William James

Nearly everything we do and nearly everything we think about involves our capacity and ability to pay attention. We pay attention to things in the world and we also pay attention to our own mental activity. Attention is how we actively engage with information. You are probably paying attention to several things right now. One of them, I hope, is this book. And as you read, your attention may shift and fluctuate. You may notice the sound of the fan, the buzz of your phone or a passing shadow. You may notice internal shifts as well. Maybe something you read reminds you of something you saw or read somewhere else.

You may also be aware that you have a curious relationship with your attention. Although you can control it enough to shift its focus, you do not have complete control. Attention also sometimes seems to be automatic.

It can seem to control you. You can follow your train of thought without thinking about it. Things can interrupt you. Stimuli and signals for the environment can seize control and redirect your attention. Thoughts and ideas from your own mind can do this as well. You're still paying attention, but the direction and location are not always up to you.

Even when you don't think you're paying attention to anything; you're still paying attention to something. You're probably paying attention to many things at any given time, though you may not be paying much attention to any of them. You may be monitoring the world for anything that may grab your full attention later: short, momentary monitoring of the incoming signals. Your brain is waiting for an important signal or stimulus, one that demands or requires more cognition, more thinking. This might sound a bit haphazard, this constant motioning and updating that might make it hard to sustain a focus on one thing for a long time. But this is also a flexible and adaptive system.

Consider this common scene: you and a friend agree to meet after work or school for coffee at Starbucks. When you arrive, you scan the store for your friend. Although it is very busy and full of people, you are able to pick out your friend almost automatically as soon as you see them. You greet your friend, place an order at the counter, and sit down. Now this being a Starbucks, the barista will take your order and your name, write your name on the cup, make the drink, and will call your name when your order is ready. In the meantime, you go back to talking with your friend, mostly unaware of all the other conversations going on. You find that you are easily able to focus on what your friend is saying and not what others are talking about, even though there is probably music playing and there are people talking. And you each have a smartphone that competes with your attention. The barista is calling out names, and you are, at best, only half aware of the names being called. You might as well be completely unaware, because you would probably not be able to remember any of the names that were called before they got to yours. You might not even be able to verify for sure if the barista even called out any names at all. That is, until they call out your name. Then your attention shifts. Quickly. Even though you are talking with your friend, and deliberately not paying attention to all the other conversions, you snap out of that and reorient to the barista. After attention shifts, you get up to get your drink and walk right back to your table and pick up the conversation where you left off and go back to ignoring all the other conversations and names being called. And just like you probably could not remember any of the names that were called out prior to yours, you probably will not remember any of the names that are called out after.

This is a common, very familiar experience. And yet, there is a lot going on in just this simple, everyday scene. Let's examine this scene more closely to see exactly what the brain and mind are doing. First, there is visual attention involved in the scanning you do as you search the store for your friend. Your attention helps you screen out unfamiliar faces and select the one you are familiar with. The same visual attention lets you focus on the menu before you order. Second, you are also using your auditory attention in the same way to ignore other conversations while you talk to your friend. But at the same time, you are still monitoring the other sounds for your own name to be called. If we were able to measure your level of attention to the conversation, we might even find that you are able to be more attentive to your friend after you get your drink rather than before, because you no longer have to devote attentional resources to listen for your name to be called. Finally, you are also paying attention to your friend on a moment-to-moment basis, to stay involved in the conversation and to understand what they say and to also pay attention to what you want to say in response. Just having a simple conversation involves monitoring at least two people speaking and switching back and forth between what you hear and what you want to say in return. It seems natural and almost automatic to most of us. Many aspects of this are automatic. They do not require attentional resources. They operate outside of conscious awareness. But there's really a lot going on behind the scenes. The task demands in this simple example are high and would be a major challenge for a machine or a computer algorithm. Imagine trying to program a computer to follow one voice (or several) while generating sentences and also monitoring for another voice to call a name. It would be a very complex program. And yet, we can do this almost automatically, without thinking about it. We rely on our attention to select, to focus, to multitask and to sustain behaviour.

Defining attention

The way attention is working in the examples above suggest that it does more than one thing. It suggests that our attention is a complex construct, one that may require complex, multifaceted definition. The way we talk about attention in our everyday language reveals something about its psychological qualities. William James wrote in 1890, that 'everyone knows what attention is'. I have the full quote at the beginning of the chapter. And like many things James wrote, he was on the mark. We all know what attention is, at least in a general sense. Indeed, we use the term attention today to mean much the same that James did in 1890. In fact, the first challenge here is to describe all the things that attention is, and some of the things that it is not.

Let's unpack James' quote a bit and pull out some of the main topics of attention because those things are what this chapter is about. James writes that attention is taking possession 'of one out of what seem several simultaneously possible objects or trains of thought'. The important thing here is 'one out of several'. One way to think about attention is to look at our ability to select one thing out of several. We'll call this *selective attention* and define that as the cognitive resources needed to select something in the environment or in our memory that we want to process further or think about.[14] Selective attention is what happens when you are talking to a friend at a busy Starbucks. There are many other sights and sounds and you need to mostly ignore those things (other people's conversations, other sounds, drink orders being called out) so that you can selectively attend to the person you are talking to. Selective attention is also what happens when you are reading this book, if you need to pay more attention to what you are reading and less attention to what is going on around you. You select the stimulus you need to process. You select the thing that you need to accomplish a goal.

James also mentions 'trains of thought', which is a conceptual metaphor that evokes an image of one thought being connected to the next and sort of pulling the whole collection of thoughts along. We'll call this aspect of attention *sustained attention* and define it as the result of cognitive processes that are required to remain engaged in the same thought or task from one moment to the next. We attend to features and aspects of things in the environment as a way to keep our attention locked in. Otherwise,

---

14 As we saw in Chapter 4 and will see later when we discuss memory, it's not always clear how and where to define things that are in the environment and things that are in our memory. Because we have to process incoming information and blend that with memory, the line between what's 'out there' and what's 'in the mind' is very fuzzy.

our minds will start to wander and begin to look for other things to pay attention to.

James also talks about attention as involving conscious concentration. We'll call this *focused attention* and we'll define this as the process of relying on conscious effort to keep something in your attention. Notice that this is related to selective attention, but it is not exactly the same thing. And it's related to sustained attention, but it is not exactly the same thing either. Finally, James talks about 'consciousness', which implies the active nature of attention and hints at the relationship between attention and *working memory*, which is the kind of short-term memory that helps us to process the things that are right in front of us. I will have more to say about working memory in Chapter 6 but for now, we can describe it as your most immediate form of memory. There is one other aspect of attention that is in our Starbucks example that is not captured in the James quote and that is your ability to monitor the world or the scene for things that capture your attention for more processing. We'll call this final idea *attention capture*.

For the next part of the chapter, I'm going to talk about how psychologists came to understand how attention works. Once we understand how attention has been conceptualised, we can then explore the modern understanding of attention. This will help us to answer questions like 'how can I improve my ability to pay attention?' Or 'can I learn to be a better multitasker?' and 'how can I stop my mind from wandering?'

Selective Attention

The modern study of attention began, like many areas of the study of psychology, with military funding.[15] In the first half of the twentieth century, the United Kingdom was at war with Germany, first in World War I and then two decades later in World War II. It was during that second war that the two countries began to look to air power as one of the most critical aspects of military strategy. Both sides of the conflict wanted to improve air

---

15 I'm always amazed at how much of our modern cognitive science age came from military funding. IQ tests, personality tests, computers, research on attention, and research on teamwork all got their start in the military. The modern smartphone is made possible with broadband, GPS, cellular networks, digital computers . . . all of which are direct products of military spending. Even the existence of the internet itself owes a debt to military spending.

power and the ability of their human pilots. Experimental psychology was still a new science, but it was seen as a way to try to understand more about the limits of human performance. Psychologists in the US had been working with the military on assessment and testing but it was work with British pilots that really showed how psychology could be used to understand human ability and performance. For example, Fredric Bartlett, known for research on memory and thinking (Bartlett, 1932) founded one of the first laboratories for applied psychology at Cambridge University, which was dedicated to using the new psychological science to help the Allies win the war. It was also during this time that Alan Turing was working with his machine to break German encryption, a remarkable story, with a tragic end, that has been told in books and film, most recently the 2014 movie *The Imitation Game* with Benedict Cumberbatch as Turing.

When the war ended, the research did not. Two psychologists, Colin Cherry and Donald Broadbent, continued to work on problems of applied psychology and aviation. One major challenge for pilots both in wartime and peacetime is having to pay attention to many different signals. This is, of course, a challenge for everyone, but seems to be particularly pronounced in pilots who have to fly the plane, monitor dozens of instrument panels, as well as carry on conversations with their co-pilot, staff and ground crew. Most pilots can do this reasonably well. One of the things that Cherry discovered was that even as pilots were focusing on the flight and a dialogue with their co-pilot, they had little difficulty switching to the ground crew or other radio chatter when it was important. Just like the example I began with, where you barely notice the barista calling names until they call your name. Cherry referred to this as the *cocktail party phenomenon.*

We're all familiar with this effect, even if we do not attend cocktail parties. It has to do with a situation in which you are fully engaged in a conversation with someone. The key here is *fully engaged.* Oh sure, we can also carry on a conversation in a half-hearted way and tune out from it, not really listening. We all do that sometime, our mind wandering to our smartphones or what we want to make for dinner. But when you are fully engaged in a conversation, you tend to focus on the person and the topic of the conversation. The cocktail party phenomenon arises from a specific scenario in which you are fully engaged in a conversation and someone else who is not part of the conversation says your name and your attention momentarily flags and switches to the person who spoke your name. It's as if, even during an intense conversation where you are giving your full attention, there was still some residual part of your attentional system monitoring the environment for important information. And your name is about as important as it gets.

We can continue to describe scenes like this but in order to study it psychologically you need to design a controlled experiment. And that is what Cherry did. He invented a psychological task known as the *dichotic listening task*. This is designed not only to mimic the cocktail party scenario, but also to push it to an extreme in order to isolate the effect. It's called *dichotic* because it involves listening to two different messages, one in each ear. Wearing a set of headphones, one spoken message is played to the right ear, and another spoken message is played to the left ear. And in this task, you are supposed to pay attention to one ear only. Now that's tricky enough to manage, but in order to ensure that the people in the task are really using all of their attention, they are asked to shadow the message in one ear. That is, the person tries to repeat back everything they hear in one ear, as soon as they hear it. This is a very demanding task. Imagine listening to two different audiobooks at the same time, one on each ear, and trying to repeat back one of them as you listen. You would not be able to pay attention to anything in the other ear at all, because all of your attention and processing capacity would be focused on one ear and not the other. And while they are shadowing the message in one ear, the message in the other ear plays along and is essentially ignored because all of the person's attention is devoted to the shadowing task. It's like a very intense cocktail party in your ears.

What Cherry did next is the key to the task. While people were shadowing the message that was being played in one year, using nearly all of their available cognitive and attentional capacity, they were thought to be completely ignoring the message in the unattended ear. Cherry, however, was interested in whether or not they could glean any meaning or semantic content despite ignoring the message. At the end of the experiment, they were asked about the content of the message in the unattended ear. This might seem unfair and unnatural. After all, if you are carrying on a conversation with a friend at Starbucks (as in the example at the start of the chapter), you would be surprised if your friend then asked you to answer questions about what someone else was talking about, or what names the barista was calling out before they got to yours. But that's the whole point of the dichotic listening task. We want to know what gets through from an unattended source. This is critical, because if we want to understand how your attentional system is able to select and focus on some cues in the environment at the expense of others, we need to determine what the system uses to make the selection.

The early research discovered that attention selection tended to be directed by what psychologists call *low-level features*. These are features which are very close to the physical aspects of a signal but rarely if ever carry any meaning. In sound, this would mean location in space, pitch, volume and tone. In vision this would mean light, movements and location. In Cherry's dichotic listening studies, people listen to a message in one ear and shadow it and essentially ignore the other message. They do not have much of a choice, because the shadowing task is so demanding. After a few minutes, the experimenter asks the listeners questions about the message they did not pay attention to. That is, if they were shadowing the message in their right ear, the message they heard in their left ear.

Researchers discovered that people cannot understand much of what is being said in the unattended ear. They cannot repeat back any of the message. They cannot detect or report on any individual words. People cannot detect a switch from one language to another. And they cannot even tell the difference between words and non-words. In the classic version of this experiment, very little meaning is able to be detected or understood. Some aspects of the message do seem to get through. For example, people can tell if the unattended message is speech or non-speech sounds (tones or noise). People can also accurately report if the voice in the unattended ear switched from a man's voice to a woman's voice. It seems as if the attention filtering mechanism is operating with just enough input to be able to gather information about low level, perceptual information like tone, pitch and loudness. Those are the things you would need if you wanted to pick a message out of many and follow it. Those are the things that you need if you want to pay attention to your friend at Starbucks and not get distracted by people around you talking. In other words, your brain pays a little attention about the physical aspects of the sound, just enough to be able to pick out the right message but not so much that the content of these unattended messages would compete for more extensive attention and processing.

Bottlenecks in the attention stream

Research with dichotic listening suggested that there were limits to how much a person can pay attention to at one time. This surely lines up with your own intuitions. It probably seems self-evident. The challenge is designing a theory or a model of how these limitations work in the mind and in the brain. As with many theories, there is often a metaphor that inspired the model. In this case, the metaphor is of a bottleneck. The neck of a bottle is the narrowest part of a bottle. The bottleneck acts to restrict the flow of whatever liquid is in the bottle and only lets a small amount in or out. An attentional bottleneck, then, is a mechanism that would act to restrict the flow of information (rather than liquid) into the brain.[16] One possibility is that this functional bottleneck is somewhere in the input stream. This bottleneck restricts the brain from being able to process everything at once. This bottleneck limits what you can hear and understand. Indirectly, this bottleneck is serving the same role as the receptive fields and complex cells of the visual system that I described in Chapter 4. Recall that vision is a process of computationally abstracting information and losing some of the detail. Attentional bottlenecks in listening do the same thing, albeit in a very different way. But the goal seems to be the same: to extract rapidly the information that is needed at the expense of truthful detail.

If there is a functional bottleneck for processing information, where would that bottleneck be? Early in processing or later? Donald Broadbent designed a general-purpose, early selection model of attention that, although incomplete, helped to drive research in this area for decades. According to his model, auditory attention is an informational bottleneck that only lets in some information, so that you process only what you need to. There is unlimited capacity at a low level in the system. Everything is processed. All the sound hitting your ear is initially present. Just like all the visual features are present in the retinal image when I described vision in Chapter 4. Then the bottleneck, early in the system, lets in information as needed for later processing.

---

16 Again, we're back to using a 'fluid metaphor'. So many of the theories we've discussed are built on top of this deeper metaphor. I find that it can be a challenge even to discuss cognition and thinking without resorting to this metaphor. Let's face it: while the brain is clearly an electro-chemical network, the mind often still seems to be hydraulic.

This model sounds pretty simple. But when you think about this idea for a few minutes, you realise that it is far more complicated. First of all, how do you know what information is important? How can you let in the important information before you know what it is? How can you keep out the unimportant information before you know that it's unimportant? The very idea of a bottleneck is not as simple as it seems. We need a system that can solve the problem of letting in too much but without introducing another problem of having to know what the information is ahead of time. In vision, the cognitive system solves this problem with the early abstraction of visual features. In auditory attention, according to this theory, the bottleneck handles the information in an analogous way. The bottleneck can be *switched* by low-level physical attributes. It's not a passive bottleneck, but an active switch to limit some information at the expense of other information. The auditory attention system cues in on these features and attributes in order to make quick decisions about what is important.

In order to include some information and exclude other information, the bottleneck model needs to be able to detect and select simple features that will operate the attentional switch. Your attention system needs basic, primitive features that may not have any meaning in and of themselves, but are predictive of meaning, objects and things in the world. The connection between feature and object is similar to what we saw in Chapter 4, with visual perception. Lines and edges may not carry meaning themselves, but they are predictive of meaning. Lines and edges exist in the visual stream because light is reflected by objects in structured and reliable ways. Lines and edges exist in the visual stream because they were most likely caused by an object. A model of selective attention for sounds depends on the same connection between sound and object.

What are some of the candidates for these primitive features? *Location* is one. If you hear a loud sound on your right, you will quickly orient your attention to where the sound is. You might turn your ears toward that sound. A non-human animal would do the same, even without the same level of knowledge or language. Cats are especially good at this. They might seem to be sleeping but can move their ears to orient to a sound. Location is content-free. Location is a purely physical feature. On its own it does not mean anything, and so it can be used to select information in the early selection bottleneck model. Location is also connected to the existence of objects.

A chirping bird will generate sound from the same area, obviously because it is in the same area. A person talking will generate their speaking sounds from the same area. The object (the bird or person) is making the sounds, and so location, which is inherently a low-level and physical feature, can be used to gather information about an object even before the object is identified.

You can also detect differences in *pitch* or *tone* without ascribing any meaning to the feature. High-pitched frequencies sound different from low-pitched frequencies because sound waves push the air in physically different ways. High-pitched sounds have high-frequency waves, which means that the energy in the wave is packed tighter and closer together. Low-pitched sounds have low-frequency waves. That is a purely physical feature. There does not need to be any meaning connected to frequency. But objects and things create sounds according to the characteristics of those objects. A small dog makes a higher pitched bark than a big dog, because when they bark, they push the air in different ways. The big dog has a big head, a big mouth. These push more air at lower frequencies than small dogs with smaller heads. Pitch is generated by the physics of the object. It's perceived by the physics of the ear. And so, it makes an ideal candidate for low-level auditory attention.

Loudness is another low-level feature, having to do with the amplitude or size of the sound waves. More sound energy from the source of the sound means bigger, taller sound waves. If you yell at someone, you push more air with your voice, using more energy, than if you simply speak. And because sound energy dissipates over space and time, objects that are closer will tend to sound louder because more of that energy reaches your ears. And just like pitch, we can sense and perceive information about loudness. Loudness is also a candidate for a low-level, primitive auditory feature.

A final candidate is *timbre* which corresponds to the quality of a sound. Most sounds are not pure sounds. Most objects do not generate pure tones with a sine wave. Sounds waves are complex combinations of many waves with complex, convoluted shapes. As with location, pitch and loudness, timbre (or sound quality) arises because of the shape of the object that generates the sound. For example, a cello and a piano can play the same note, from the same location and at the same loudness, but they will still sound very different. The piano makes sound a different way than a cello does. It pushes the air around in different ways, creating a wave with a different shape and different complexity. Your ear and attention can detect this feature before any meaning is attached to it.

The biggest challenge in describing how selective attention works is to figure out how we can select something in the environment to pay attention to before we know what that thing is. This might sound like a trivial problem, but like most trivial-sounding problems, it's far more complex than how it initially seems. Features are part of objects, but in order to put the features with the objects, you have to know what the object is. But, of course, in order to know what the object is, you need features. You can see where this is going. It's kind of an unsolvable, circular argument. It's part of a larger psychological/philosophical problem called the *binding problem*. The binding problem is a problem in understanding how we bind together what we see and hear in a way that reflects actual objects in the world. This problem crops up in lots of places. We saw it in Chapter 4 when we discussed visual perception. It's cropping up now with selective attention. In order to survive and thrive, we need to be able to pay attention to things in the world. Things that are needed for survival, things that are needed for communication with others, and things that are needed for pleasure. The bottleneck model I'm describing offers one solution, though it's not the only model of this process. And as we'll eventually see, it is not a complete description.

The first problem we have to solve with selective attention, is what do to about all the information that is present initially. Just as in vision, in which we argued that all the visual information strikes the retina, we have to contend with having all the sounds that are in your environment strike your ear. According to the bottleneck theory we have an unlimited capacity low in the system. In this case, 'low' means closer to the ear. Our ears have no voice but sense everything. All the sound in the environment strikes your ear at the same time. The ear processes the sound wave and is able to perceive information about pitch, loudness and the like. I do not have the space in this book to go into the details about how the ear does this, but the essence is the same as what we saw in vision. There are receptors that respond to the physical energy and convert it to neural information. All that processed information reaches your ear and is sent on to the primary auditory cortex in the temporal lobe.

Let's examine this in the context of selecting a channel of information to pay attention to, as in the dichotic listening task. You're listening to one message in one ear and another message in another ear. I described the task earlier; you have to repeat the information in one ear. You selectively attend to that channel and ignore the other. Then we might ask about what you were able to pay attention to in the unattended channel. So right off the bat, we have a problem to solve. And some information about how we solve it. We have to pay attention to one channel. One ear. The easiest way to do this is with that location cue. We can trip the attentional switch to only really process stuff from one ear. Easy. But some information does make it in. People can tell if a voice in the unattended ear switched from a man's voice to woman's voice. In this case, the information about timbre and pitch is still being monitored by the unattended ear, which is ready to process the information further and possibly even trip the switch. If you were shadowing a man's voice in the right ear and the man's voice switched to the left ear, your attention would follow those low-level cues. The timbre and pitch of the voice would capture your attention and would switch the locus of your attention and conceptual processing to the other ear. You might not even know that you started following the other ear. In principle, you've made a mistake. You started following the wrong ear and did not even realise it. But in practice, your attentional system made a reasonable inference. You were paying attention to a man's voice at the beginning, and even when the location of the voice switched, you would follow the voice to another location. This is a practical adaptation. People move around in their environments and change location. But sounds coming from one source usually do not suddenly change their timbre.

This example shows the bottleneck and early selection system in action. Even as you pay attention to one ear and shut out the other, low-level information like pitch and timbre is still being processed. It's still activating feature detectors. And because of that, it is able to 'trip the switch'. In this case, when paying attention to someone's voice, it is going to be more important, more useful, more adaptive to incorporate things like pitch and timbre. After all, you don't actually identify a person by where their voice is coming from. You identify a person by what their voice sounds like. So that information trips the switch.

This bottleneck theory works by assuming that there is this unlimited capacity at first, and that features can be used to trip the switch and select for any message to pay attention to and to process further. As with vision, this all happens before you comprehend the meaning. This happens before you know what you are listening to. This happens as a consequence of the cognitive architecture and the way the system is built.

The bottleneck model, as described by Broadbent, is a decent model. It describes much of what happens in this kind of selective attention. And it accounts for a lot of the data that has been observed. There's one glaring problem, though. It doesn't explain the cocktail party phenomenon, which means that this model does not explain the Starbucks example I began with. Why not? According to this model, the capacity is unlimited low in the system, but it is restricted before any meaning is processed. The only way to maintain or switch attention is for the system to latch on to low-level, perceptual, meaning-free features. But if only low-level features can trip the switch, how can your name get through to pull your attention away? There seems to be no way for the model to process meaning in a way to account for this. The model may not be wrong, but the model does not seem to be complete.

In order to provide a complete account of selective attention, and to account for the cocktail party phenomenon (feel free to call this the Starbucks effect if you like, or even the coffee-shop effect) we need to consider a later selection model. In other words, we need to move the attentional bottleneck.

Early and late selection

The idea of an attentional bottle neck makes intuitive sense. We just cannot process everything, and we need to have some way to select the things we want to pay attention to. But the bottleneck idea does not quite work for everything. No matter how focused you are on something, carrying on a conversation for example or playing a video game, you can almost always be pulled out of that focus if someone calls your name. Low-level perceptual features can help to limit attention and restrict what we pay attention to, but the bottleneck may not be that low in the processing stream. Some information must be getting through in order for things like your personal name to get through.

Why is this important? We deal with multiple inputs all the time and we get things out of order sometimes. You can see that in the little mistakes that we all make. Whether we're reading or writing or just catching up on

Twitter or Instagram we are also monitoring the environment for other important signals. Suppose you are immersed in reading the news online and your partner walks in and asks you a question. For a split second, you might be confused about who said what. You might even answer their question incorrectly based on what you were reading rather than what they ask. Sometimes information is mixed because we're not really able to screen out one channel at the expense of another. If most unattended information was filtered out early, before it was processed for meaning, we would not be able to pay attention to a conversation and have our attention be diverted by having someone calling our name. We would miss our order at Starbucks. Although early, pre-processed, perceptual features are important for a model of attention we also need a way to take in more information.

Revised versions of the bottleneck theory, most notably the work of Anne Treisman, a British psychologist who did most of her work at Princeton, suggested that these low-level physical features are still important in selective attention, but the selection is not at the location of perception but rather the selection is based on a response to information in the environment. In other words, you don't screen things out. Instead, information is processed but you screen things from being selected and acted on once they are in the system. You see, one of the biggest problems with Broadbent's theory of attention is that it assumes we select information to let in. If you never let the information in, it is no longer available for processing. But we know even from our simple Starbucks example that information that you are not paying attention to *does* get into our cognitive systems. In the Starbucks example, you would barely hear the barista calling out names. You would not register anything and hearing the names being read out would not interfere with your ability to concentrate on your conversation. Until you hear your name, of course.

How might this late-selection theory work? Anne Treisman claimed that the selection and the bottleneck is much later in processing. Her work suggested that we hear everything and most of that information makes it into our brains and minds, and then we choose the information to respond to. According to Treisman this information is all available in the minds for a brief period until you process it further. This idea builds in the early-selection bottleneck theory by emphasising the importance of primitive features. It adds to the idea that the selection and filter are later in processing and thus susceptible to more cognitive processing. If you assume that some words and concepts have special importance you can then assume that those words have a lower threshold of activation. Your name, for example, is important. It's arguably the most important word there is for anyone. It's probably one of the first words you heard. As a result, it can be said to have a low threshold of activation and you will recognise and respond to it even when the information coming from the environment is quiet, confused, unattended or otherwise degraded. It's as if your ability to recognise your name is being run by a name-detector module that monitors the environment for any trace of your name. It's on high alert all the time. As soon as it detects the slightest sign of your name, it sounds the alarm, your attention is grabbed, and you orient toward that signal.

Treisman's model explains our coffee shop scenario, but there are other things it does also. In the classic dichotic listening experiments, people pay attention to one channel (i.e. information presented in one ear) and they miss the meaning of anything in the other ear. In the original studies, the unattended ear was completely separated from the attended ear. That is, there was no reason to combine them. But some studies were done with information that conformed to a narrative in one ear and another narrative in the other ear, and if the narrative switched from the left ear to the right ear, most research participants would switch along with it. If you heard a story in the left ear and the plot switched mid-sentence to the right ear, the only way to make any sense of this would be to follow the story. This just does not work with Broadbent's early selection model because his theory assumes that unattended information was prevented from entering. Treisman's model assumes that information from both the attended and the unattended channels were able to enter into the cognitive workspace and then you select the information that makes sense.

The most effective model seems to be one that does let you tune into these low-level, meaning-free features but still lets a lot of the information in. Features activate your auditory receptors, then send their cascade of activation to the temporal lobe of the brain where word meaning is processed. This information spreads to other areas of the temporal, frontal and even parietal lobes, where meaning is processed. The information you are attending to, that you are processing, that you are working on, activates itself. It activates what is already active and so you are able to maintain your attention. The information that you are not attending to just floats away and decays. But it is in the system. It's there if you need it for a brief time. It can activate key concepts, like your name. It can even override the attention to low-level features and make you switch your focus without realising. This is beneficial for survival. This is also the source of many of our frustrations with paying attention. It's not easy and many things still get in and interfere. Broadbent's theory might be incomplete, but to be completely honest, I often wish I could select a channel and screen out all the extraneous information from even being processed at all. There are times when I wish the bottleneck was earlier and narrower.

Capacity

Many years ago, when my children were in that middle ground between preschool and middle school, we went to an indoor recreation centre in our town. We live in Southern Ontario, and sometimes in the winter months, it can really be daunting with the ice, the snow, slush, etc. Kids like to go out and play in the snow, but they also get bored and four or five months of bad winter weather can be monotonous. Parents look for indoor things to do. Indoor recreation centres like waterparks, trampoline parks, gyms, and bowling alleys fill that need for physical activity and are popular everywhere.

This particular indoor recreation facility had things like bowling, rock climbing walls, and an indoor miniature golf course called 'glow golf'. Maybe you've seen something like this. This is an indoor mini golf putting game where you try to get your ball through obstacles and into the cup. Added to that are black lights, which cause some things like the balls, the sticks, and some of your clothing to glow. It's otherwise kind of dark and has a little bit of that haunted house vibe. This could be fun, we thought. Or at least, this will be fun for two kids below ten, they would enjoy the game, laugh at their glowing clothes, and we would have fun watching them have fun. Importantly, it is not an especially demanding game and therefore should not require a lot of attention.

It was fun for the first ten minutes. However, it was also very popular and there were a lot of other families and kids. Our kids were still pretty young, and so it took time to get through each station. There were obstacles. That's just the way miniature golf is. If you want to enjoy it, it can take a few minutes for each hole. But behind us was a larger group of older kids, right on our heels. They were probably twelve to fourteen years old. They were much faster, better at the game, louder, more restless. Because of the way a game like miniature golf works, they were right behind us. Every time we finished playing a hole, they were standing right behind waiting, running around, pretending to hit each other with golf clubs, and making the experience much more stressful than it should have been. I offered to let them play through, but theirs was such a large group that, even though some of them were right behind us, others in their party were still playing and they did not want to pass us up. So, for the next 45 minutes, as I was trying to play with my girls, and keep an eye on them at their young age, there was a group of rambunctious twelve-year-olds right at my heels. It took an enormous amount of concentration to *ignore* their activity and try to focus on enjoying my game. By the end of the glow golf game, I felt completely exhausted. Not from the glow golf. Glow golf is not in and of itself an exhausting exercise. I was exhausted from all of the cognitive control I had to exert to ignore the loud kids behind me. My attentional capacity was taxed.

Cognitive control, which is a form of attention, takes effort. We seem to have a limited capacity to pay attention, to ignore paying attention and to deploy our cognitive resources on a problem. We see this in so many other ways.

Have you ever felt exhausted after taking an exam? Or working on a difficult task at work that required a lot of attention? Your capacity was probably diminished by the end of the day. Maybe you even reached a point where you were performing at a suboptimal level. You probably felt too exhausted and it might even have affected your judgements. It's at this period of diminished or exhausted capacity that mistakes are made, and tempers might even flare. This is where people miss important information. This is where we might fail to restrain ourselves. This is where we are less able to control our attention and behaviour. This is where people get angry and short-tempered.

How can we avoid this? One possibility is to switch what you are doing and let yourself recharge. When I'm feeling exhausted from one thing, teaching for example, I notice that my capacity is not affected for some other things. I can still get in my car, listen to music, and drive home. I might be tired of thinking about a research problem, but it does not seem to affect my ability to start cooking dinner. When I switch to another activity, I feel like my capacity is no longer drained. And because I cannot resist using the 'fluid metaphor', I am just going to go ahead and say it: One attentional resource pool has dried up, but the other one is still full.

The psychologist Lee Brooks, who worked most of his career at McMaster university in Hamilton Ontario, was an influential researcher in cognitive psychology. Lee died in 2010, but he was one of my academic and intellectual heroes. More than just about anyone else, his research really affected the way I think about how the mind works. One of his earliest studies showed just how these attentional pools work and how we keep them separate. The experiment and the task are clever and creative, so I want to spend a few paragraphs explaining it. You will thank me for this.

In Brooks' experiment (Brooks, 1967), subjects are asked to do either a visual imagery task or a verbal imagery task. In visual imagery, they are shown a shape and asked to memorise it well enough that they can imagine it later, to visualise and inspect the image in their mind's eye. In one experiment, the shape was a large outline letter with an asterisk on one side. They were told to memorise this and then imagine the asterisk tracing around the letter in their mind's eye. So, picture a large letter 'F', drawn as an outline with an asterisk at the bottom left and then imagine the asterisk tracing around the letter (see the figure for an example). Sometimes the asterisk would be on the outer edge of the figure and other times it would be on the inner edge. At each juncture, subjects would indicate YES or NO as to whether or not the asterisk was at an inner or outer edge. Pretty easy right? It is. You can do this now, if you like. Look at the image below, commit it to memory and then try to do the imagine task and when the asterisk is at the extreme edge, imagine YES and when it is at the inner junction, imagine NO. For this shape, it would be YES, YES, YES, YES, NO, NO, YES, YES, NO, YES.

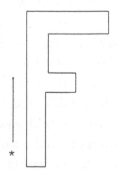

5.1: an example of a letter imagery task

The verbal imagery was just as easy. Instead of a picture, subjects were asked to memorise a sentence. For example, subjects would read the sentence like 'A bird in the hand is not in the bush' and commit it to memory well enough that they could repeat it back using their inner voice. And while you are imagining the sentence with your inner voice, you can also imagine a YES or NO answer to the question of whether or not each successive word is a noun or not. This is also pretty easy to do. And for this sentence the answer is NO, YES, NO, NO, YES, NO, NO, NO, NO, YES.

These tasks are easy to do in the mind, but if you try each one of them, you might find that it's easy enough to memorise and image the sentence, but also just a bit easier to imagine saying YES or NO to the visual imagery. Brooks reasoned that we're often using two attentional pools. We use visual attention and verbal attention and if a task draws from the same pool, there is a cost to performance. If the tasks draw from two separate pools, then there is little or no cost.

Brooks tested this in a clever way. In addition to being asked to learn either a visual or a verbal image, subjects were also asked to respond in one of two ways: a visual-spatial response or a verbal response. Visual spatial responding involved pointing to a 'Y' or an 'N' that were scattered on a sheet, and you had to look and point in different areas. The verbal response was just to say 'YES' or 'NO' out loud while the experimenter listened to you.

Subjects could be in one of four conditions. A visual–visual condition in which they memorised the image and pointed, a visual–verbal condition in which they memorised the image and spoke, a verbal–visual condition in which they memorised a sentence and pointed, and a verbal–verbal condition in which they memorised a sentence and spoke. Brooks predicted, and found, that when the imagery and the response were drawing from the same attentional pool, people were slower and made more mistakes than when the imagery and the task were drawing from different pools. Visualisation and perception, it seems, compete for the same neural responses. When they draw from the same pool, it can drain that pool more quickly.

This is why you can walk and talk, or drive while you carry on a conversation, or cook dinner while you watch a video. These are two different tasks with very different perceptual and attentional demands. This is why you cannot pay attention to two conversations very well or listen to music with vocals while you try to write.

It seems we can multitask to a point, but that it's far more difficult when the multitasking involves the sharing of attentional resources. Let's talk about that idea, multitasking, in more detail. We all think we can multitask, and we do multitask by virtue of the way our cognitive systems evolved. But as Brooks' work shows, there are costs and these costs will increase to the extent that the things you are trying to multitask are similar.

Multitasking

One way to think about the need for selective attention and multitasking is to take a moment and consider in explicit detail everything that is going on around you while you are reading this book. For example, you are probably holding the book or eBook reader in your hand. The book has a certain amount of weight to it, and you can pay attention to that (or not). But even if you do not pay explicit attention to that, your motor system does. It has to adjust your grip and support so that you are able to hold the book and not lose your grip. Your motor system reacts *automatically* to take into account this weight so that you do not drop it. It also adjusts the support so that you do not pull it up too much. There is considerable neural and cognitive computation going on in just that once aspect, but you are probably not aware of it unless you take the time to concentrate on that one thing. This delicate push and pull with grip, grasp, gravity and the book's mass is a set of behaviours that mostly goes on outside of conscious attention. And that's just one of the many things that's happening.

There may also be a variety of ambient noises around you that are not important for your reading. Perhaps there's a fan running, a cat purring, a kettle boiling, or a car running by. You hear them slightly but do not attend further. Or perhaps you are listening to music that is perceptible but not distracting. Again, you need to screen these things out. Maybe you're at the beach and there are people around talking, kids laughing. Perceptible and part of your awareness, but only at a low level.

You carry out sensory and neural activity that is part of your reading, but not part of *what* you're reading and so you are not even aware that these computations happen. You are moving your eyes over the words. You maintain a focus on a single line. You recognise words. You construct a mental model of the sentence. You are reminded of other ideas and you read. Are you aware of the concepts and ideas as they are activated and put together? Or are you only aware of the product. You turn pages. You breathe. You take a drink, you swallow. Your mind wanders. You bring it back.

Except for thinking about the ideas in the book, a lot of these actions and behaviours go on without your even being aware. You are doing several things at once and only really paying attention to one of them. In other words, you are multitasking. You are concentrating on one thing (reading) while a lot of the other things happen automatically. The more you focus on one thing, the more those other things are shuffled over to 'auto pilot'. Or, as James put it in the quote at the beginning of the chapter, attention: *'implies withdrawal from some things in order to deal effectively with others'.* When reading a book, we withdraw from paying attention to most of these other behaviours and actions so that we can deal effectively with the ideas in what we are reading.

Notice one key aspect of this scenario. You are doing several things at once, but only one of them counts as the thing you are paying attention to. The others are going on automatically. That means that if you did try to pay attention to some of these behaviours or processes, we would have to withdraw from the main task of reading. Try to concentrate on your grip and the act of holding the book and how it feels in your hand at the same time that you try to focus on what you are reading. It might be possible, but what you should find is that your ability to concentrate on the book is reduced. You are multitasking, and that *'implies withdrawal from some things in order to deal effectively with others'.*

This raises a question. If we have to withdraw from something in order to multitask, then why do we multitask? Multitasking is natural, adaptive and unavoidable. We really do need to be able to do more than one thing at a time. We need to pay attention to where we are walking but at the same time, we need to be aware of obstacles, of others, of our own thoughts. Our cognitive systems evolved to do this. It's impossible to avoid because there are so many signals. But many of the signals, most of them, are not important. There is a lot of noise in the world, both literally and metaphorically. We can't process it all. So, we multitask and switch around, and there is a cost. Even that cost is beneficial, because it implies that we can concentrate on one thing and withdraw (somewhat) on what does not matter. This is adaptive. This is good for survival. This is not always good for thinking.

One of the biggest challenges to thinking (and learning and paying attention) is multitasking. We multitask all the time, and we are fortunate to have brains and minds that have evolved to allow us to process many different streams of information and to switch quickly between them. The problem is, most of us think we are much better at multitasking than we really are.

I teach at a large university, and I lecture in front of a class that can be anywhere from seventy-five to a hundred and fifty students. One of the things I notice when I look out into the class is that most students are taking notes on their laptop or a device like a laptop, such a Surface Pro or an iPad pro. In addition, they often have their smartphone nearby. The most common configuration is to see a student taking notes on their laptop and occasionally looking at their smartphone which is placed to the immediate left or right of their laptop. Attending a lecture is a combination of looking at the lecturer, looking at the slides on the projector or white board, looking at your laptop, typing some notes, looking at your phone, looking back at the lecture, looking back at what you typed, then glancing at your phone. Each 'look' is a *withdrawal from some things in order to deal effectively with others.* You have to wonder how effective this is, though, with all that switching going on.

This is not a criticism of my students, by the way. In fact, I think my students are usually very attentive and engaged in the lecture. But this switching around is just the way things are. And of course, it is not just students. If anything, professors are worse. If I look around the room at a department meeting, a faculty meeting or a lecture where there might be a lot of professors in attendance, many of them also have a laptop and/or a smartphone. I see many professors only half listening to a presentation while they manage emails on their laptop or check things on their smartphone. I am guilty of doing the same thing. I often take notes on an iPad Pro and flip back and forth between email, notetaking and occasionally Twitter.

And it's not just in academic settings either; people do this when watching television. Some popular shows practically depend on the fact that people watch and engage in social media at the same time. People watched *Game of Thrones* with one hand on the remote and another on their smartphone so they could post and tweet their reactions. People 'live tweet' a presidential debate. I've coached my daughter's softball teams for years and I notice parents on their phones extensively even as they watch the game. Many of the same parents who stare at their phones also requested that I put a 'no phones' policy in place for players on the bench. But the girls are on their phones anyway. Even me, the coach, will pull out my phone to update the score on an app that pushes the score to an app on parent's phones so that they can see the score of the game, making then all the more reliant on their phones during the game.

The situation is much more serious on the roads. The next time you are driving somewhere or on the bus and you are at a red light or a junction, look around at the other drivers beside you. Chances are, many of them will be sitting in their cars looking down at their phones. This is often despite the steep fines for being on your phone while driving. It's been against the law where I live (in Ontario) for a few years now, but I still see it. Conservatively, I would say about 75 per cent of all the drivers I see stopped at the light are looking at their phones. This is not safe. I have even noticed people looking down at their phones while they are driving. Looking up at the road and down at the phone then back up at the road again. This is not safe! We know it is not safe. The drivers doing it probably know that what they are doing is not safe. And yet many of us feel that we can multitask. The problem is that we do not realise what we are missing when we are multitasking, because we are multitasking. How can we notice if we have withdrawn from some things to deal with others?

Digital multitasking, or multitasking with a smartphone, computer or device is not a new phenomenon. The means of multitasking might be new, but the process is not. As we discussed earlier, this is exactly one of the things that Cherry and Broadbent were interested in trying to study. This is what Brooks was studying. We know how it works, we know why we do it, and we know that it's usually a benefit. But the *effects* of technology and multitasking seem to make it a more pressing issue in our daily lives. Something about the nature of digital multitasking makes it seem like a new problem. But is it really a new problem? Let's look more carefully at this.

Cognitive psychology and the smartphone

The iPhone was released in 2007. When I was writing this book, in 2019, it had been on the scene for twelve years. Having been born in 1970, I am in Gen X. I did not grow up with a smartphone, but I did experience nearly the full range of mobile phone development and indeed the full range of digital communication as an adult. After growing up with a single telephone on a party line,[17] I got my first email address in university in 1988 and have used email, FTP, Lynx, UseNet, NetScape, SMS, Facebook, DMs, Slack, and many other communication forms ever since. I sometimes even still call on the phone when I absolutely have to.

When the iPhone celebrated its ten-year anniversary, it got me thinking about the relationships I've had with smartphones and mobile devices. It is funny, in some ways, to think of it as a relationship. But for many of us, it's exactly like that. We relate to our phones. We bring them with us. We think about them. They take up residence in our psyche. Maybe it is not quite as well developed as a relationship with a person. Maybe more like a relationship with a pet.

Of course, as with pets, I remember many of them. I can picture them as if they were real relationships. The first one was solid but simple and only made calls, there was no camera and no messaging. That was followed by one or two forgettable flip phones and a phone that began to incorporate media player options but never really lived up to its promise. I remember these, but not that well. I did not have a strong relationship with any of them.

But then came the iPhone, and everything changed. I started really loving my phone. It became a relationship, in part, because the phone enabled better communication with people. It took photos. It let me communicate with friends and colleagues in many more ways than the flip phone. The relationship was built on top of the same framework that formed the basis of most real, actual relationships. Maybe even more so, because the iPhone was the connection to everyone. And unlike a telephone or a computer, I could bring the phone everywhere. I could connect to my friends and indeed the entire internet, anytime, anywhere and everywhere.

As a result of this, and in combination with the advances in industrial design that produced more beautiful phones, it got harder for people to be away from their phones. I know for me it became a much bigger habit after

---

17 In rural areas with limited connectivity, and up through the 1970s and 1980s, a 'party line' was a single telephone line, literally a 'line' as in a 'wire', that several households would share. We had different phone numbers, but we were all on the same, single wire. That meant that sometimes when you wanted to make a phone call, you would pick up the phone, and there would be another person already talking on the one line that was shared between the houses on the road. This seems almost incomprehensible today.

iPhone than before. I started with a sleek, black iPhone 3GS and then a white iPhone 4S which I still regard as the pinnacle of iPhone design. In fact, I still have that phone as a backup. A move to Android since has seen me work through several models before my current smartphone. It is with me every single day, and almost all the time. Is that too much? Probably. Am I alone in this? Probably not.

Smartphones today are used for many things. There is even a chance that you are reading this on a smartphone. Most of us have one, and we probably use it for many different tasks. For example: communication, social media, taking and sharing photos, music, video, navigation, news, weather, calculator, fact checker and alarm clock. One thing that all of these functions have in common is that the smartphone has replaced other existing ways of accomplishing the same tasks. Fifteen years ago, I would have had a phone, a CD player, a camera, a calculator, a road atlas, a newspaper, an encyclopaedia, and a television to accomplish the same things. Now, I have one small smartphone that does all of this faster and better and does so much more than that. It's no wonder that we are fixated. It's replaced everything else and that attention is all on one small thing. That was original idea for the iPhone, one device to do many things. Not unlike 'the One Ring', the smart phone has become the one device to rule them all. Does it rule us also?

The psychological cost of having a smartphone

For many people, the device is always with them. Just look around in any public area: it's full of people on their phones. As such, the smartphone starts to become part of who we are. This ubiquity could have psychological consequences. And there have been several studies looking at the costs. Here are two that piqued my interest.

A few years ago, Cary Stothart carried out an interesting experiment in which research subjects were asked to engage in an attention monitoring task that required people to see a series of numbers flashed on the screen while they had their phones with them (Stothart, Mitchum & Yehnert, 2015). This task, called the Sustained Attention Response Task (SART) is common in attention research. In the SART task, subjects are asked to press a key on the keyboard as quickly as they can every time a number flashes, except when the number '3' flashes. When '3' is flashed, they have to withhold their keypress.

This task requires sustained attention, vigilance and also inhibitory control. It's not difficult, but it's not able to be done automatically either. If you were multitasking, for example, you might not do very well.

Subjects performed the task twice. For both sessions, they were asked to keep their phones with them. The first time through was to see how well they would perform as a baseline. In general, people performed well. Then they were asked to do it again. It should be at least as easy the second time through, but the subjects did not realise that when they had signed up and consented to the experiment, they had provided their mobile phone number to the experimenters. This meant that the researchers could text or call them while they were doing the experiment. Which is exactly what happened next.

On the second session, one third of the participants received random text notifications at an ongoing but variable rate while they did the task. Imagine taking an exam or watching a show carefully and your phone starts going off. Another third of the subjects received a random call to their phone which is even more stressful. Calls are more insistent, the notifications pulse with more urgency, and because most people text or message, a call often has more importance. The other third of the subjects proceeded as they did in the first session, with no additional interference beyond what might be expected. Subjects in the control condition performed at the same level on the second session as they did on the first session. So far so good. But the group of subjects who received random notifications (text or call) made significantly more errors on the task during the second session. And not surprisingly, calls affected performance more than text messages did. In other words, there was a real cost to getting a notification. Each buzz or notification distracted the person just a bit, but enough to reduce performance. This study was carried out in 2015 when it was not uncommon to have your phone make a noise or a vibration. Silent settings have become more common now. So, put your phone on 'silent'? Maybe not. For one thing, Stothard's study only analysed subjects who did not interact with their phones despite the text. That is, a notification might pop up, but they were not supposed to look. So maybe just thinking about the notification was enough. That was their claim. One way to test this further would be to have people be near their phones and not even have the notifications. Is your phone nearby? Do you find yourself being distracted?

A paper recently published by Adrian Ward and colleagues (Ward, Duke, Gneezy & Bos, 2017) seems to suggest that just having your phone near you can interfere with some cognitive processing. In their study, they asked 448 undergraduate volunteers to come into their lab and participate in a series of psychological tests. The volunteers were randomly assigned to one of three conditions: *desk, pocket/bag* or *other room*. People in the *other room* condition left all of their belongings in the lobby before entering the testing room. People in the *desk* condition left most of their belongings in the lobby but took their phones into the testing room and were instructed to place their phones face down on the desk. Volunteers in the *pocket/bag* condition carried all of their belongings into the testing room with them and kept their phones wherever they naturally would (usually pocket or bag). Phones were kept on silent.

The participants in all three groups then engaged in a test of working memory and executive functioning called the 'operation span' task (or O-Span). In the O-Span, people have to work out the solution to some basic maths tests and keep track of letters at the same time. It requires a lot of attention and a lot of cognitive capacity. It's not going to let you multitask. If you were doing something else that drew from the same general pool, you might have to withdraw some attention from this task. They also asked their volunteers to perform a Raven's progressive matrices task which is a test of fluid intelligence.[18] The results were striking. In both cases, having the phone near you significantly reduced your performance on these tasks. If the phone was on your desk, just sitting there, and not making any sounds, you were still slightly affected.

A second study found that people who were more dependent were affected more by the phone. This is not good news for someone like me, who seems to have his phone always nearby. They write, '*Those who depend most on their devices suffer the most from their salience, and benefit the most from their absence.*' One of my graduate students is currently running a replication study of this work and (assuming it replicated) we expect to look in greater detail at just how and why the smartphone is having this effect.

---

18 Fluid intelligence is a term that refers to problem solving and reasoning intelligence, as opposed to knowledge and the retrieval of facts.

The implication of this work is that you might not be thinking about your phone that much, but maybe just glancing over at it from time to time out of habit more than anything else. But this habit is still a cognitive switch and attentional switching always has a cost. Stealing a quick glance at your phone might just be enough to throw you off your game. It might just be enough to make you forget one of the list of letters you were supposed to remember. It might make you miss an important piece of information in the Raven's task or cost you seconds on a sustained attention test.

The long-range implications are even more serious. Even if you are diligently not texting and making sure to stay off email and social media while driving, you might still have your phone on the console. Maybe you use it to stream music while you drive or listen to a podcast. Stealing a quick glance at your phone, even out of habit, might just be enough to make you take your eyes off the road for the split second that a pedestrian, also stealing a quick glance at their phone, steps onto the crossing . . .

With the inherent risks, one wonders if smartphones are a smart idea? I suppose they are. The implication is just that phones, like any distraction, will come at a cost. The locus of this cost is not so much the phone, but it is a consequence of the way our minds work. Like so many other things we've discussed, our minds are adaptive and allow us to create, navigate, solve and think. The cognitive architecture that allows us to do this so well sometimes gets the better of us and also produces mistakes. The mistakes are the cost of doing business. We often cannot stop from making them unless we take special measures. That is what many people do with their phones. We find ways to stop making these mistakes.

Despite the many uses for these devices, I still wonder how helpful they really are, for me at least. When I am writing or working, I often turn the WiFi off, or use a blocking app like Freedom to reduce digital distractions. But I still have my phone sitting right on the desk and I catch myself looking at it. I did it just now, as I wrote that sentence. There is a cost to that. I tell students to put their phones on silent and in their bag during an exam. There is a cost to that. I tell students to put them on the desk on silent mode during lectures. There is a cost to that. When driving, I might have the phone in view because I use it to play music and navigate with Google Maps. I use Android Auto to maximise display and mute notifications and distractions. But there is a cost to that.

It is a love—hate relationship for many people. One of the reasons I still have my ancient iPhone 4S is because it is slow and has no email/social media apps. I'll take it with me on a camping trip or hike so that I have weather, maps, phone and text, but nothing else: it seems less distracting. Though it seems weird to have to own a second phone to keep me from being distracted by my 'real' one. Many of us spend hundreds on a smartphone and more per month for a data usage plan and at the same time, have to develop strategies to avoid using the device. It must seem a strange paradox of modern life that we pay to use something that we have to work hard to avoid using.

CHAPTER 6

Memory: An Imperfect Process

In Chapter 4, I argued that you can't always trust your senses. That's because we do not perceive the world exactly as it is. What we think we see and hear is a reconstruction of what was there moments earlier. What we think we perceive directly is actually processed information. And information processing takes time. Information processing implies a process. We lose some information in the process. And it also implies that even as we think we are living in the present, we are living in a reconstruction of the immediate past. By the time we register what is in front of us and realise what we are looking at, what we are hearing, and what we are smelling, the world has moved on. In Chapter 5, I suggested that we learn to multitask as a way to withdraw from things that may not be directly important and commit our cognitive resources to what may be more relevant and important. As with the gaps in perception, this means that we are going to miss some information.

In other words, our perceptual and attentional systems have developed and evolved a strategy to deal with the constant and continual barrage of information in the world by just missing some of it. We withdraw from following some streams and we redirect attention to follow other streams where they might lead. We'll never be able to follow both of them at the same time. We fill in some gaps here and reconstruct some others there.

But there is an advantage to the way we process information and to the fact that we lose information. Living in a reconstructed world presents us with a beneficial trade-off. When we look at a visual scene, for example, we can use our existing knowledge to fill in some of the details of that scene.

When we recall and use this existing knowledge, we essentially perceive and process anew just the information that is most likely to be useful for making predictions and decisions. We reinforce the association and strengthen the links between the memory and what we are perceiving. Does that mean we fail to take in some bits of information and that we might fail to notice some things? Of course, but that occasional failure is the cost of the efficiency we have evolved. In general, we have evolved and adapted to this trade-off of missing some of the details that we expected to see and processing and attending to those things that might be new, novel or noteworthy.

You may have already realised the paradox in what I'm arguing for. If we are using our knowledge to fill in some of the details, where does that knowledge and information come from? It's part of our memory, of course. We use our existing knowledge and memory to fill in many of the details from perception that we might not encode fully when we look at a scene or listen to something. This is efficient and a benefit to our brains, our mind and cognition in general because we don't have to work on perceiving everything in a familiar scene all the time. Great, right? Yes, sometimes this is great, but not always. If you are using your knowledge to fill in some of the gaps and details, then you are not actually seeing what is right in front of you. Rather, you are seeing a blend of what is in front of you and what is in your memory. And that means that sometimes you might think you see something or hear something that isn't really there.

This creates a little bit of a problem. When you use your memory to fill in the details, you are inferring what is *supposed* to be there, filling in things that are usually there, but maybe not always there. This is a probabilistic process. Your brain makes a guess about what you are seeing. It may be imperfect, but it usually works for us. But a guess is still a guess. Most of the time, your brain makes a good guess. But sometimes, your memory is going to make the wrong guess. In that case, you will make an error. Good guesses and bad guesses (errors) come from the same place. They are both the result of your memory trying to fill in the details so that you can concentrate on the task at hand.

Guessing About the Past and Present

How do these memory guesses work? Let's look at a simple example. Think about the street in front of your house, the road in front of your home.

Now suppose there is almost always a blue car parked across the street from where you live. You walk out of your door and there it is. You come home from work and there it is. The car probably belongs to someone who lives near you. Maybe they live across the street. Maybe there is nowhere else to park and that is their preferred place. If you see it there all day, it might mean that the owner is near all day. Now before I go on, I want to point out how we are already filling in details from our memory for an event that did not even happen. This is just imaginary. I asked you to imagine the car. And still, we have no trouble making assumptions and inferences about the car, its owner, and their life. This is how our memory works. It fills in background and details whether we want it to or not. It predicts possibilities whether we want it to or not. Memory is working continuously with perception and attention to try to bring order and sense to the world.

Returning to our example, if you quickly glanced outside, you would expect to see the car at certain and predictable times. If the car is always there in the afternoon, for example, your memory makes a note and generates an expectation that the car will be there in the afternoon. Reliable and consistent. This expectation might be strong enough that if, by some chance, the car was *not* there, your memory might fill in those details anyway. That is, you might think you saw the car even if it was not there. Your memory would be filling in the wrong details. Or you could argue that it's filling in the right details and the wrong time. You might not even notice if it was missing one day. Or, if you tried to remember the scene later in the day, you might not even be able to say for sure if you saw it that day or not. Your memory knows that it is usually there, and so the safe, certain guess is on thinking that it was there.

For the most part, that kind of filling in the details and guessing is a good thing. It means you don't need to note specifically that the car is there each day. This is good, because you can think about other things. This is an example of your brain matching patterns and making good, realisable predictions. And if you think it was there when it was not, it's probably not going to cause you any problems. Sure, thinking the car was there when it was not is technically a memory error, but it is an innocuous one. Using your memory to make guesses about the present is very efficient and unavoidable, but it also means you make unavoidable errors from time to time.

The science and psychology of memory is the science and psychology of errors. There are many kinds of memory errors. We all know that there are times that our memories let us down. We often tend to focus on our faulty memory in terms of forgetting things. We forget people's names. We forget where we put our smartphone. We forget to call the dentist. We forget birthdays and anniversaries. Forgetting things is only one kind of memory error, but it is the kind of error that we are often most aware of. Forgetting something means facing the consequences of that error. It's easy to notice when we forget something, because there's something missing, or someone reminds us of what we forgot.

But there are other kinds of memory errors. Errors that are much less noticeable. Errors that are sometimes closely related to the predictions and inferences that our memory helps us with. These kinds of false memories are more difficult to notice, more pernicious, because they are working in conjunction with the natural tendency to use our knowledge and memories to fill in the details. It is hard to tell the difference between a false memory in which we fill in the wrong details and this natural process of just filling in the details. Furthermore, this kind of false memory goes unnoticed because it often doesn't matter. But sometimes, of course, it does matter.

Memory Is an Untrustworthy Companion

Let me tell you a story that I remember from my teenage years. Most of us have many memories of various kinds for personal events and things that happened to us. I will describe these kinds of memories as 'episodic memory' in the next two chapters. For most of us, these memories take the form of stories that we recall and tell. Good stories are interesting. Good stories have interesting details, interesting characters, and an interesting plot. We tell them to people and to ourselves to inform, explain and entertain. Stories, of course, are a form of fiction.

My story is one of many stories that I have told and retold. I am telling it here because it has to do with memory errors. It is a story about a story. I don't remember the exact date,[19] but it was sometime between 1986 and 1988. I can say this not because of a personal memory for the date, but because of

---

19 One of the most frustrating things about memory is that we seem to remember some things vividly, like a story, but other things not at all. Curious that what I do remember is a subjective experience, but not something concrete like the date or even the year. I remember how I felt or how I think I felt but can't even agree on my age or the year.

a factual form of memory that is called 'semantic memory'. I had just learned to drive in 1986 when I turned sixteen. I did not own a car until many years later and the events unfolded before I went away to college. So general, factual knowledge lets me conclude that it was between 1986 and 1988 when I graduated high school. I will have more to say about the different kinds of long-term memory in Chapter 7 and Chapter 8. For now, let's just assume that personal memory is not always the same as general factual memory.

I was driving my dad's Ford Bronco. The Ford Broncos in the early '80s were built like pickup trucks and had mirrors on the sides that extended outward half an arm's length, like a truck. These mirrors were attached to the vehicle with a metal hinge so that they could be moved back and forth to make adjustments. I was driving with a friend. And as we were driving, I must have drifted a little bit to the right-hand side of the road and the passenger side mirror made contact with something, probably a telegraph pole or a sign, which caused it to swing inward on its hinge and strike the glass of the passenger window. It broke the window and sprayed shattered glass on my passenger (who was not injured).

Although it was good that no one was injured, it was not good that I broke the passenger window of my dad's Ford Bronco. Not a serious collision but still something that was a problem. Something that was going to have to be repaired. Something that was going to make the car less drivable. Something that I was going to have to pay for. Something that was also going to cause him to be upset. As soon as I got home, I told him about the accident. I told him that I must have clipped a telegraph pole or sign with the passenger side mirror. But why had I veered too close to the pole? I had no clear answer, but I thought that I must have veered to the right in order to avoid an oncoming car in the other lane. That seemed to offer an explanation for why I couldn't avoid having this accident.

The problem is that I am not entirely sure if my recollection was true. It explained my actions and it fitted into a story. But to this day I cannot recall exactly what happened 2–3 seconds prior. Although I am not sure if this part of my story is true, I am also not sure the story was a lie either. I did not purposefully tell an untruth to deflect blame. I just did not remember either way. I just needed to have an explanation and I filled in some details. I created a coherent story using what I knew had happened and what I thought might have happened.

I do not think I remembered all the details then. And I surely do not remember them now. But I do remember telling him the story, and so my memory of the event is that it must have been true. But I also remember wondering if he would believe my explanation, so then I remember that the event may not have played out that way, and that I lied to my father. Was the story true? Did I forget something? Did I embellish? Did I lie? I do not remember. I have a memory of feeling sure about the story but also a memory of feeling unsure of the story. Neither one of them is a strong memory. Now, all I remember is the story about the story.

We probably all have a lot of memories like this. Memories for an event that is partially remembered. What we end up remembering now is the *act of remembering itself* more so than the event. Each time we remember the event (or remember the remembering of the event) it is subject to some possible change. If we observe some new detail, make a new interpretation, or even blend some current information with the memory, we will store this blend as part of the memory for the next time we recall it. Each recollection brings with it the possibility and probability of a change. In short, our memories are far from stable. Our memories are imperfect to begin with and they are constantly changing.

I am getting ahead of myself, though.

Before we discuss any more about false memories, we should discuss memory and thinking in general. In order to understand how and why these false memories come around, and why we trust our memories despite their fluidity and malleability, it is worth exploring a little bit about how memory works, how psychologists have discovered how memory works, and how the brain operates to create the experience we call memory. This chapter will cover how we access our memories, how memories influence our thinking, and how false memories may not be that different from the 'true' memories. Then, in the next two chapters, I'll cover the cognitive psychology of short-term memory, long-term memory, and knowledge.

If you want to improve your thinking, you need to be able to understand how your memory works. If you want to learn to discern the accuracy of your intuitions, you need to be able to understand how memory influences behaviour. If you want to make judgements and decisions with confidence and accuracy, then you need to be able to know how memory works, when to trust it, and when not to trust it. Thinking is all about memory and good thinking depends on a good memory.

What Is Memory, Anyway?

We tend to think of memory as being a record of the past, as being mostly past oriented. But one of the most surprising things about memory is that it is not really about the past. Your memory is about the future. Memory is how we use things that happened in the past to inform what we do now in the present and what we will do in the future. Memory may have the veneer of the past, but it also has the functionality of predicting the future. Functionally, it would not really make much sense to have a memory system that just let you relive the past for its own sake. We remember the past so that we can make sense of the present and predict future outcomes and events.

So, what is memory? That seems like a pretty straightforward question. Until you try to answer it of course. Is memory a kind of internal storage system for something that happened in the past? Is memory a record of everything that is in your mind and brain? Is memory a process that allows you to move forwards and backwards in time? Is memory a conscious mental process or is it an unconscious mental process? Are your memories stored and retrieved or are they experienced and re-experienced? The answer to these questions is, yes . . . sort of.

We tend to regard memory as a mental process that is analogous to a file drawer or a computer hard drive: you have an experience and then you file the experience away in memory where it can be retrieved later when needed. But that's not really how memory works. Everything we experience, we experience through our memories. This is even true for direct observation, because as soon as you perceive something, the thing that is in front of you has already changed. It takes a few milliseconds for the light energy to travel from your eye to your primary visual cortex to your temporal lobe where you can recognise the object. At which point, what you perceive visually is not exactly what is in front of you, but a reconstructed memory of what was in front of you a few milliseconds ago. The same is true of hearing; by the time the sound has travelled from someone else's lips to your ear, it is already gone from the world forever. All you have left is your memory of the sound. And as we discussed earlier, what you perceive is a blending of what is actually there in the world with what you think is there, based on what you remember and based on what you know. The neural processes of perception and memory overlap. Your memory is a form of reconstructed perception. And because your perception is enhanced by memory, that implies that even your perception is a reconstructed perception too.

Before I get too deep in the weeds of questioning objective reality, let's define memory as the process of recognising that a pattern of neural activation that is occurring now is similar to a pattern that occurred before. The recognition doesn't have to be overt and explicit, all that matters is you behave the same way, or that your brain is able to treat the correspondence between the current pattern of activation and the previous pattern of activation as being similar. That's memory.

Memory and Thinking

To some degree, the process of thinking itself is nothing more than the process of using your memory. When we learn something, we get better at recognising the similarity between scenarios or events that happened in the past and scenarios and events that are happening now. Learning is a process of strengthening a connection between what you know (your memory) and what you don't know. We use prior evidence to make decisions, to solve problems, and to make judgements about the world. We use what we know and what we think we know to guide our behaviours. Thinking is using your memory to make decisions, plans and judgements.

One important way that we use our memories to guide thinking is to assess new situations for risk. We then use our assessments of risk to plan actions and to guide behaviour. We find ourselves in both risky and non-risky situations all the time. But one of the inherent aspects of risk is uncertainty. If we are familiar with a situation and a circumstance, this implies that we have some memory for a previous similar situation. We can use that memory and the sense of familiarity to reduce the uncertainty associated with a new situation. We can recognise the risk and the possible outcomes because we have faced them before. When we do not recognise the risk, the situation or the outcome, that will increase the uncertainty. If you are in a situation that is dangerous to you, but you don't recognise the risk because you don't have any relevant memories available, you may behave inappropriately. If you don't have access to a relevant memory, you could end up putting yourself in danger. Even worse, in an attempt to reduce uncertainty, we might recall the wrong thing. Or sometimes, when faced with a new situation, we remember similar situations from the past, but they may not be the right ones to guide behaviour. In these cases, our attempts to use the past to guide behaviour can have negative consequences.

This was the case in early 2020, when world leaders, doctors and citizens first faced the threat of the novel coronavirus that caused the COVID-19 outbreak. Although the virus was new, some aspects of the virus and the reaction to it were similar to viruses, pandemics and crises that had occurred in the past. Many of us behaved in ways that indicated we were trying to reduce the uncertainty of the situation by doing what we did in previous crises. But this sometimes turned out to be a mistake. The virus was too new and too different. One of the most critical mistakes was made by the mayor of New York City, Bill de Blasio, in early March 2020. Perhaps remembering the approach that previous leaders took after the September 11 terrorist attacks, in which people were encouraged not to stay home but to go to the theatre and restaurants, de Blasio made an infamous announcement on Twitter. He wrote:

Bill de Blasio ✓
@BilldeBlasio

Since I'm encouraging New Yorkers to go on with your lives + get out on the town despite Coronavirus, I thought I would offer some suggestions. Here's the first: thru Thurs 3/5 go see "The Traitor" @FilmLinc. If "The Wire" was a true story + set in Italy, it would be this film.

8:16 PM · Mar 2, 2020 · Twitter for iPhone

2.2K Retweets **1.7K** Likes

Figure 6.1: A tweet from New York mayor Bill de Blasio.

A month later, thousands of New Yorkers were dying every single day from this virus. De Blasio made a mistake, but his mistake was understandable. He was behaving like he remembered the advice from 9/11 or like he was remembering the 'Keep Calm and Carry On' posters from the World War II era in the UK.[20] He was falling back on the idea, in memory, that people should not let the current adversity keep them from doing normal things. He was falling back on the memory that we should keep calm and carry on. But the coronavirus outbreak of 2020 was not like these previous situations. And as we later found out, this was really bad advice. The appeal to the familiar, though understandable and all but unavoidable, was not the right decision.

This reliance on memory to assess risk is a necessary part of what we do and how we interact with the world. This tendency is present early in life and we use our memories to determine risk from a young age. To use a straightforward example, a young toddler just beginning to walk does not know that a kitchen stove could be dangerous. How could they? In fact, they might think it is a good thing to be around. After all, they might see their parents or caregivers working at the stove when they are cooking breakfast, lunch or dinner and their memory would store a connection between their caregiver, the stove and the food. That's probably a good memory. In addition, a stove is just the right height to be interesting to a child that is learning to walk. It may have knobs on the front and a door that pulls down to open up the oven. From the perspective of the toddler, a stove is an exciting thing that is associated with lots of good things. And so, it is probably natural to think that a child wants to get close to it.

However, there is obviously a danger associated with it too. If the child has no memory or no prior experience with being burned at the stove, then they have no reason to treat it as a dangerous, risky situation. In fact, all those good memories might increase the tendency to want to be near the stove. Thankfully, most children do not have this kind of negative experience. Most children have not been burned by the stove. In the absence of a direct negative experience in memory, we still want to make

---

20 An example of memory elaboration it its own right, the 'Keep Calm' posters were printed in the late 1930s but never officially displayed. What we remember are the recreations based on a poster discovered in a bookstore in the 2000s. We just think they were part of wartime Britain, but they were not.

sure children avoid the stove. Instead, we usually create a mildly negative experience for them to remember. We give them a reason to remember that the stove is not so good. The way to do that is for the caregiver or the parent to issue a stern warning. They yell or startle the child when they get close to the stove. 'Stop! Don't touch the stove!' you might say. You might even feel bad saying it because it might scare the child. But what you are trying to do is create a memory for the child to ensure that there is a higher probability of a mildly unpleasant event associated with the stove. In this case the unpleasant event is the scolding, rather than a more serious burn. We want the child to remember being scolded (rather than scalded) so that when they approach the stove again, their judgement of the stove is that it is predictably dangerous. We want that memory of being scolded to be *available* to the child so that it is there when they need it. We want to reduce the uncertainty and to increase the salience of the memory for the negative event.

Having something available in your memory allows you to make a quick judgement. The child has an available memory of having been scolded. This memory is available because it comes to mind quickly and easily. And because it's now available in the mind, the child can quickly and reliably judge the stove as something to be cautious around. In this way, the availability of the memory is a useful and helpful heuristic. This availability provides a way for memory to help the child and to make quick judgements. This heuristic can, in a very real way, help to ensure survival. Memory alters behaviour so that the curious child will behave appropriately cautious around the stove, despite having never been burned. All of this happens in an imperceptible flash. Available memory can be a life saver.

Both examples, De Blasio's mistake about COVID and the toddler's available memory that helps them avoid the stove, highlight a mental chain of events that links perception, attention, memory and cognition. That is, the present is coloured by the past. The effect is automatic and unavoidable. You cannot ignore the cascade of neural activation that occurs when a current situation, object or event brings to mind previous, similar knowledge. Seeing the stove calls to mind instantly the mildly negative association with a past event. The automatic availability is crucial in this case. If the risk of danger is high and the time course for injury is very quick, we would do well to make the judgement very quickly if not instantaneously.

What better way to quickly assess and diagnose a situation than to rely on the first, strongest and most salient memory? If the memory seems strong, it must be the right memory to use. If the memory is recalled quickly and without effort, and is available in our minds, then it is understood to be a fast and reliable cognitive shortcut that we can use to inform our decisions and judgements about the world. And so, we use that information to make a judgement, an inference, or an adjustment to our behaviour. Problem solved, right?

Maybe not. As with so many of the concepts I have been discussing in this book, the cognitive processes and the cognitive architecture that lead us through the world and help us to behave adaptively are the very same cognitive processes and architecture that sometimes lead us astray. That's probably what happened to Bill de Blasio (and many others who might have been too casual about COVID early on). Available memory that seems to offer guidance did not result in the right decision. When these cognitive shortcuts help, we call them heuristics. When these cognitive short cuts hurt or cause us to make mistakes, we call them biases. Helpful heuristics and harmful biases are two outcomes of the same underlying mental process. And that process is the tendency to use memory to guide our assessments and perception of the world. Let's look at the heuristics and biases in more detail.

The availability heuristic

Daniel Kahneman and Amos Tversky spent years studying how people use their memories and knowledge to make decisions and assessments of risk. They refer to this tendency as the *availability heuristic*. As I have described, the availability heuristic refers to the tendency to base a judgement on what is most available in our memory. Or, more precisely, it is the tendency to make use of the ease with which a memory can be made available. In that way, our memories can fool us because sometimes the wrong information comes to mind. Sometimes, the wrong information is surprisingly easy to bring to mind. Other times, the wrong information is available because we don't have a complete picture. If inaccurate information comes to mind, then we might make a mistake. These mistakes may not always be costly, but they are inaccurate, and they reflect the tension between what something is (objectively) and what we think it is (subjectively). When these things are in tension, it's all but impossible *not* to resolve the tension in favour of what we think something is. We almost always favour our subjective interpretation of events. This results in a bias. We experience this bias because some things that come to mind easily and are therefore available may not accurately reflect reality. Let's look at a recent example. Something that illustrates the nature of the availability bias.

In 2014, a number of NFL players in the United States were accused, arrested or otherwise implicated in high-profile cases of domestic violence. This was covered extensively and was a top story even for people who might otherwise not follow American football. The specific case that many people remember involved Ray Rice, who was a running back with the Baltimore Ravens. Mr. Rice was seen on a security video assaulting his fiancée (now wife) in a lift. The video was truly awful. Rice was seen hitting her and dragging her limp body out of the lift. It was a brutal display. Most people remember Ray Rice for this and not his career on the football field (which was very good until then). In fact, Rice lost his position on the team and no longer plays football because of it. At the time, there was a lot of media coverage, wide availability of photographs and video, it was discussed in social media and on the news. People were talking about it a lot, reading about it, and seeing the video. In a survey that was conducted around that time by the popular sports and statistics website fivethirtyeight.com, nearly 70 per cent of Americans endorsed the opinion that 'the NFL has a troubling epidemic of domestic violence'.

The problem is that the objective evidence did not exactly support that conclusion. I don't mean to minimise domestic violence or Rice's assault. Domestic violence is a problem. It's existence in the NFL and otherwise is troubling. But it was not accurate to say that in 2014 the NFL, in particular, had a troubling epidemic of domestic violence. In fact, in 2014 NFL players had a noticeably lower rate of arrests for nearly all crimes, including domestic violence, compared with the general population. Ben Morris, a statistician who works with fivethirtyeight.com, used Bureau of Justice statistics to calculate the arrests per 100,000 people for all men aged twenty-five to twenty-nine, which is the age range for most NFL players. What he found is that the rates were lower for driving under the influence, for nondomestic assault, for drug-related offenses, for disorderly conduct, for sex offenders, for theft and yes, for domestic violence. It seems that NFL players, at least in 2014, were much less likely than average to be arrested for just about anything, including domestic assault.[21]

So why did people think that the NFL had such a troubling epidemic of domestic violence? The reason probably seems obvious in light of the availability heuristic I've been discussing. The survey was conducted when the topic of domestic violence, and especially Ray Rice, was on everyone's mind. People were talking about it, posting about it, writing about it and thinking about it. It was very available. And so, when someone was asked in a simple poll if they agreed or disagreed with the premise that the NFL had a troubling epidemic of domestic violence, and they'd just been reading about Ray Rice and a handful of other cases, it was going to influence their judgement. The first thing that comes to mind is a series of high-profile domestic violence cases. What doesn't come to mind is domestic violence in other professions, like law enforcement or medicine. What doesn't come to mind is all of the NFL players who were not arrested for domestic violence that year. What certainly doesn't come to mind are the arrest rates per 100,000 men aged twenty-five to twenty-nine as per the Bureau of Justice statistics. That is, we have an available memory for a video of a graphic assault that was shown over and over again but we do not have any available memory (or any memory) for crime statistics.

This is an example of the availability heuristic at work. Since we don't generally have access to accurate information like arrest rates or the base rate of occurrence, we tend to rely on what we do have access to, which is our memory for something. This available memory is usually all we have. It is fast and usually reliable. The same process that allows us to learn to avoid a stove when we are young by having an available memory of being scolded also leads us to make errors in judgement. We use what we have.

Daniel Kahneman gives many straightforward examples in his book, *Thinking Fast and Slow* (Kahneman, 2011). The most well-known example is from an early study by Kahneman and Amos Tversky. They asked a group of research participants questions like 'suppose one samples a word of three letters or more at random from an English text. Is it more likely

---

21 You can read the original report here, and it shows that although NFL players are arrested at a lower rate than the general population, domestic violence is still a problem and is the leading cause of arrest if players are arrested for something https://fivethirtyeight.com/features/the-rate-of-domestic-violence-arrests-among-nfl-players/

that the word starts with the letter R or that R is the third letter?' They found that people agreed that it was more likely that the word sampled at random would start with the letter R. But that's not true. In fact, you are far more likely to randomly select a word with R in the third position. If you do not believe me, look at this paragraph so far (up to here). There are fifteen words with R in the third position. The word 'word' even has R in the third position. And there are only four words that begin with R. Words with R in the third position are more common than words that begin with R and so if you choose one at random it is much more likely to be a word that has R in the third position.

The question arises, why do people think more words start with R when in fact more words have R in the third letter position? Why are we so easily misled? Why are we wrong? It has to do with the way our memories work. If I ask you to make a judgement of words that begin with R, you can remember of a whole list of words: rind, random, raft, riparian, river, etc. If I ask you to think of words with R in the third position, how do you do that? You have to think of them by their first letter first, and then check to see if R is in the third position, there's a lot of them of course, if you look back over the previous few sentences you will notice that the word 'word', 'more', and 'first' all have R in the third position, but you wouldn't think of them easily because we don't remember words that way. And so, when people try to remember words, they get a lot of words to begin with R at the beginning and not many with R in the third position. It is simply easier to recall the words with R in the first position. This ease of recall also plays a role in the judgement. When people are asked which is more likely, they go with what is available in their memory. And what is available in their memory are lots of R first letter words.

As you can see, this is like the NFL example, albeit more mundane. In both cases, some information is easily available in memory. In the Ray Rice example, the information is available because it's sensational. In the R in the first position example, the information is there because of how we remember and retrieve words. In both cases, however, the information is not really the correct information to guide the decisions. We don't have the correct information to judge domestic violence in the NFL because it's not something that we usually have access to even if we wanted it. Getting the correct information would take time. We don't have the correct information to judge words with R in the first vs third position because we can only recall words based on their first letter. In both cases, the correct information would take too long to get. In both cases the incomplete information comes to mind easily. And so, we use that information. The availability heuristic leads to this bias when the available information leads us to make the wrong decision.

Not surprisingly, this is something that advertisers, politicians and other groups or individuals who would seek to influence your behaviour take advantage of. For a long time, people in the United States had an inflated sense of the risks associated with terrorism. When the September 11 attacks occurred, it affected our thinking about the risks and likelihood of terrorism. We adjusted our behaviours and policy according to that perceived risk. But terrorism attacks in general are not very common in the United States. These kinds of attacks occur with a very low base rate. There just are not many terrorist attacks in the United States. The same is true of school shootings and mass shootings. They are tragic when they happen, and they happen far too often, but the overall base rate is still quite low. Stranger-child abductions are another example. They do happen, and when they happen, they are tragic, but the overall base rate is low. But if we are asked to judge the likelihood of these events, we would probably overestimate the risk. We would probably judge them as being more frequent than they are. This occurs because of the availability heuristic. Tragic but infrequent events are easily recalled because of their tragic nature. They are salient because they are discussed. And this information cannot be counteracted by the correct, more accurate information because we often do not know the overall base rate or true probability. All we know is what is available in memory. All we know is what we know. And if you get your news from a source that emphasises terrorism, school shootings and stranger-child abductions, it is no wonder you might overestimate the risk.

One thing I want to emphasise is that when you make these judgements, you are not exactly making a mistake. Well, yes, you are making an error in the sense that you might be overestimating the risk or likelihood of an event, but to some degree you are behaving exactly the way your mind is designed to behave. When you make predictions and judgements, you are taking advantage of what you see, what you know, and what you remember. Most of the time, this leads us in the right direction. We avoid dangerous things, even if they have only been dangerous a few times. It is evolutionarily adaptive to take limited information and make quick judgements based on what comes to mind. It only seems like an error when we look at the bigger picture.

Representativeness

Sometimes, memory affects the way we perceive and interact with people and this can lead us to generalise from concepts and to form stereotypes. For example, in the mid-2010s[22] in the United States, at the end of Barack Obama's presidency, there were some cases of police being attacked by protestors. In a few cases, these protests were in response to the shooting of unarmed civilians by police in Ferguson, Missouri. When that initial event led to some crimes against law enforcement officers, some media outlets in the US referred to this as a 'War on Cops'. Notice the framing. To describe something as a 'War' should call to mind and make available a concept that involves something that demands a long and concentrated effort. It should call to mind conflict and violence. These concepts will be available. Combine that availability with frequent and sensational coverage of some current cases of police officers who were shot and what you get is a new, available memory. People will hear the phrase 'War on Cops' and remember it. They should activate memories, concepts and ideas about war. This will enhance and add to the coverage of these cases. As a result, it should not be surprising that when Americans were asked in a poll if they agreed that there was a 'war on cops', they would say 'yes'. All of the evidence is available in memory. All of the evidence is right at the top of your mind. None of the evidence is an accurate reflection of the world, however. None of the evidence reflects the overall dropping crime rate. None of the evidence reflects the fact that policing is, by and large, a safe profession. The evidence reflects *your* experience of being told something and of remembering it later. The available evidence is not really wrong: it reflects what you saw and read. It is just not the right information to answer that question. And if the available information is

---

22 When I was completing this book in 2020, the US erupted in several waves of mass protests against police brutality and protests against police treatment of Black people specifically. These were spurred on by the killing of George Floyd by the Minneapolis police but spread across the country. The 'War on Cops' framing was less prevalent, in part because of the sheer size of the protests, shifts in public opinion, and also the size of the police response.

consistent with or related to our existing stereotypes and biases, then it's even more difficult to dismiss.

Kahneman and Tversky's research can inform us on this idea also. They carried out research on how people perceived different professions and found that most of us rely on our knowledge of people as being representative of stereotypes and concepts and not about our knowledge of individuals, of base rates and of true probabilities. They called this the 'representativeness heuristic'. We know a lot about stereotypes because they rely on the same conceptual framework that our memories use. We know less about base rates because we simply do not have the experience or habit of having that information. Sometimes we even ignore base rate information when it's provided. It is as if we do not know what to do with base rate information and probabilities even if we are given the information. Instead, we treat individual cases as being representative of their category.

In one of Kahneman and Tversky's examples, research participants were asked about a description of a person and to make a judgement about that person. More often than not, they made judgement errors based on their memories for stereotypical examples. Kahneman gives the following example of a question about an individual name, 'Steve'. Assume that Steve was selected at random from a representative sample. In this case, a representative sample is one that reflects the underlying distribution of the general population. This individual, Steve, is described as follows:

> Steve is very shy and withdrawn; invariably helpful but with little interest in people or the world of reality. A meek and tidy soul, he has a need for order and structure, and a passion for detail.

Research subjects were then asked to judge whether Steve was more likely to be a librarian or a farmer. Most of the participants in the study chose 'librarian'. Why librarian? Kahneman suggests that the description of Steve's personality reminds people of someone who might work in a quiet library.[23] But librarian is not really the correct answer here.

---

23 Please do not assume that I think any or all librarians fit this stereotype. These descriptions were designed in the late 1970s and early 1980s. They were designed by Kahneman and Tversky to be deliberately stereotypical and maybe even a bit extreme. The point was that people used these stereotypes and not other information.

There are (or were when Kahneman and Tversky did the study) more people involved in farming than in library science and so technically you would be more likely to choose someone at random who is a farmer. Despite that, people use the stereotype and not the population base rates when they answer the question. People use what is in their memory and assume that one example (Steve) is representative of the stereotype or the concept. Kahneman and Tversky refer to this as the *representativeness heuristic* which means that, all things being equal, we assume that specific examples are representative or typical of the concept that is activated in memory. If Steve's description activates the concept of librarian, then it must be because Steve is, in fact, a librarian.

Representativeness is another heuristic that, like availability, is based on the contents of our memory. Like availability, representativeness can result in memories that are often correct or informative. Maybe we've met librarians and farmers and we have an idea of what each is like, but we probably do not have a base rate of occupation status. Why would we?

Kahneman and Tversky usually treat representativeness judgements as an error because, in this case, it is technically the wrong answer. It is often viewed as less than rational because people will even ignore base rate information to rely on stereotypical knowledge. But is that really an error? Is it wrong to rely on a stereotype? Suppose I said I sampled one person at random from the United States general population and described them as '*A wealthy older man, who is tall and slightly overweight with light blond hair that is longer than is common and is combed in a swoop. He is prone to boasting and exaggerations. He inspires either devoted fans or equally devoted detractors*', you would not be off base to say 'Donald Trump'. Would this be wrong? According to Kahneman and Tversky you would be wrong to overestimate the likelihood of choosing this one specific person at random. But the stereotype is too strong to avoid. More importantly, the description narrows the search in your memory to be only about people who have these attributes. Noticing the similarity to the former President is obvious and it's all but impossible not to let that influence your judgement.

Like availability, this representativeness heuristic is a double-edged sword. It helps us to make quick, useful assessments of the world, and arrive at a decision. These assessments and judgements are based on our memories, which are in turn a function of our own experience. What better way to make a judgement or a decision when we're not sure of all the information than to rely on our memory and experience?

Much of the time these quick, memory-derived judgements and decisions are correct, or correct enough for us to get by. But there are at least two problems that define the other end of this double-edged sword. First, Kahneman and Tversky's research has shown that we will rely on memory even when it conflicts with objective information about the true probabilities. We tend to trust our gut, instead of the truth. The second problem is even more troubling: Our memories are often wrong, inaccurate, distorted and incomplete. So, not only are we trusting our memories over other externally objective information, we are also trusting a very unreliable source.

I began this chapter by labelling memory as an 'untrustworthy companion' but that's only part of the problem. It's untrustworthy and yet we're prone to trust it. It's incomplete and yet we regard it as complete and accurate. Memory is responsible for intrusions, distortions and outright gaps.

Let's look at seven ways that our memories fail us. These are the 'Seven Sins of Memory'.

The Seven Sins of Memory

Human memory has always had this curious role of being the one cognitive operation that we are the most aware of, and yet on the surface, it seems unreliable. And what is an unreliable memory anyway? If you don't remember something, that may not really be an error, it may just mean that you did not pay enough attention to something. That's not exactly an error of memory. In this case, the memory may be a fairly accurate record of your failure to pay attention.

Memory is curious. The very act of remembering something even creates its own new memory, further blurring the lines between past, present and future. We need to trust our memory, but it seems untrustworthy. Memory can seem very accurate even when it is giving us the wrong information. Or it can seem to be inaccurate when it might actually be all too accurate. Memory is a record of the past that we need for the future. It's a record of the past that is changed by the present, usually without us being aware. It's how we represent stability and it's often wildly unstable. Memory is an untrustworthy partner that we have no choice but to trust.

Daniel Schacter, a cognitive neuroscientist of memory at Harvard University, laid all of this out when he published a short paper in *American Psychologist* called 'The Seven Sins of Memory' (Schacter, 1999). In this essay, he described seven ways in which the adaptive and beneficial aspects of memory all but guarantee that we will make mistakes and errors. These mistakes are predictable. Schacter also argued that these mistakes are not random, but a by-product of how our memory has evolved and of the cognitive and neural architecture that supports memory function. According to Schacter, the 'Seven Sins' are transience, absent-mindedness, blocking, misattribution, suggestibility, bias and persistence. All seven of the failures have implications for how thinking is assisted, affected by and even undermined by memory failures. But all seven can also be overcome with care and awareness.

The first two, transience and absent-mindedness, are really everyday memory failures. Information fades over time. Or sometimes information is not encoded well enough in the first place because you did not pay attention to the task at hand. In either case, what you end up with is a weak or fading memory trace. For example, you are probably aware of times when reading something – a text for a class, a reading for an assignment, or even this book – that you kind of spaced out and forgot what you just read. You're reading along, your mind wanders, you forget what you just read, and you lose your place. That kind of absent-mindedness results in a memory failure. In order to work properly, memory requires some degree of attention.

The third sin, blocking, essentially means a temporary retrieval failure or lack of access. This might be due to spreading mental activation in your memory network. As the activation spreads to many memories and concepts, they all receive some activation. If many similar memories and concepts are activated in memory, and if the level of activation is similar across several different memories, it can be difficult to determine which one is the correct memory for the occasion. If you are trying to remember the name of a famous actor or film and you have activation for many similar actors and movies, they all compete for your attention. With similar levels of activation, each may seem plausible and each activated memory can end up inhibiting the others and block the correct one from coming up. This is seen most strongly in the so-called 'tip of the tongue' phenomenon. This is where you are asked a question and you know you know the answer, but you cannot generate the answer. In many instances, you can almost *feel* yourself saying the answer. It seems that some of the information is present, including the information needed to speak the word, but the information was not activated strongly enough for conscious recollection. The subjective feeling is one in which the memory is blocked but the feeling that it's there is not. It's not uncommon, but still a very curious feeling.

We have covered three of these 'Seven Sins' that are general failures to recall something. Which means they are easy to recognise. It's easy to recognise the failure to recall something because the failure to recall is a pretty good clue that you are making a mistake. The next two, misattribution and suggestibility, are distortions and intrusions on what is already in memory. They are both errors that arise from the highly associated aspects of semantic networks. As a result, these memory errors result in the experience of actually remembering something, but it's the wrong thing. It is more difficult to notice these errors because you think you're right.

In the case of misattribution, an error comes about when a person remembers the fact correctly but may not be able to remember the correct source. If someone tells you a story, even if it's not true, you might easily recall the contents of the story later on and believe that it's true. We see examples of this in politics and news all the time. Suppose a political leader speaks a falsehood (a lie) during an interview. This happens a lot. The news media then covers the story of the leader telling this falsehood. The news story is recounted and shared many times in the news. The overall effect is that the lie is repeated. A repeated lie is fertile ground for possible misattribution. Why? Well, the repeated exposure will make the facts more available. If you remember some things about the lie, but not the original source (e.g. that it was a *story* about a lie) then you end up having a strong memory for the contents of the lie but the wrong attribution. When this happens, you end up believing your memory and believing the lie.

Many politicians and leaders are able to control the narrative and spread false information by putting out falsehoods and getting the news to cover it. The more outrageous the better, because it's more likely to be covered by a lot of different media. President Trump excelled at this. He was able to generate a lot of coverage for misleading and false statements that were then covered widely. This wide coverage, even by news stories attempting to discredit the original statements, often paradoxically spread the lie further and increased the chances that people would falsely remember that the original lie was true. The error comes as a result of misattribution because the information was recalled, but the source of that information was not correctly recalled. This is an error, and one that is perfectly understandable given the way memory works.

Another one of the seven sins, suggestibility, is related to misattribution and our tendency to recall some of the information we have in memory, but to recall it incorrectly. To say that our memory is suggestible means that we often update our memory for past events based on a current description. If you are remembering an event, let's say the time you cracked the window on your parents' car, as in my story earlier, and someone suggests a new detail while you are remembering that event, the new detail that was suggested can also become part of the memory. There does not even need to be another person doing the suggesting. You can suggest new interpretations to yourself and those can become part of the memory too. In my story about cracking the mirror on the Ford Bronco, that's exactly what happened. I did not have all the information encoded in the initial memory because the event was over so quickly. As I tried to make sense of this event, I suggested some possible explanations and they became part of the memory too. Memory is easy to change, easy to adapt, and easy to shift around. It's not hard to fool yourself.

If you can fool yourself, then others can fool you too. Others can trick you into doubting your own memory and perceptions. This can be seen most vividly in the idea of 'gaslighting'. Gaslighting, if you are not familiar with the term, is a form of psychological manipulation that takes its name from a great old British suspense movie called *Gaslight*. In the movie (which I will not spoil but you should try to watch) a husband tries to convince his wife that she is going crazy by repeatedly denying what she sees and perceives. She begins to doubt her own memory. She tries resolve the conflict between what she thinks she saw with what she is told that she saw by accepting that she must be losing her mind. Many reporters and writers have accused President Trump of 'gaslighting' Americans.[24] The administration had

---

24 I am trying not to be too partisan here. This kind of influencing and guiding of public perception is a tool that many political leaders use. I'm focusing on President Trump for two reasons. First, Trump seemed especially committed to this style of political communication. Second, it's the availability heuristic. That is what was in the news when I was writing and probably shaped my perceptions and biases also.

been accused of trying to convince Americans that what they saw and remembered (whether it was the size of the crowd at his inauguration or the president's initial response to the COVID-19 pandemic) is not what happened. The eventual goal was that the new information, provided by the Trump administration, would become part of the person's memory.

Suggestibility and misattribution can be difficult to deal with because they are not easy to recognise. On the one hand, we do want our memory to be able to take new information into account and update our understanding of events. That's the underpinning of basic learning. And the way our memory works is that the contents of our thoughts are a mix of what we are seeing now (perception) with reconstructions and representations of similar things that we saw earlier (memory). This usually happens in a system we call 'working memory'. I'll describe how this working memory system operates in the next chapter, but what this means is that we are always enhancing our perception of the world with information from memory. We are always updating our memory with information from the world. The blending of these things reinforces our long-term memory. If some of the new information is false, it can still be blended with the information from memory and form the new memory. This is often unavoidable and it's a function of how our memories work. And, as we'll see in the sixth of the seven sins, we tend to trust our memories which can just make things worse.

It's bad enough that we have suggestible memories that are prone to mix up attributions. What makes it worse is the sixth sin of memory: bias. According to Schachter and other memory researchers, we show a strong bias to assume that what we remember reflects things and events that actually happened. In other words, we have a bias to trust our own memories. And with good reason. We use our memories to understand things, to learn things, and even to perceive the world that is in front of us. Everything we do and everything we have done is filtered through our memory. Everything we know, everything we have a name for, and everything we think about is a product of our memories. So, we need to trust our memories, or everything would fall apart. If a false memory or an error arises from a feeling of remembrance that comes from having a densely connected network or related thoughts and ideas, it's perfectly reasonable for us to assume that we are remembering something that happened rather than just responding to a state of activation that does not reflect the past.

In short, if we remember it, we tend to believe it because there's a memory for it. If it didn't happen, then we assume that there would not be a memory there. We kind of treat the memory as a proof that something did happen.

There's one more sin that Schacter mentions. It's slightly different from the others, which are all concerned with memory errors, forgetting and confusion. The seventh sin of memory is persistence. As much as we want most of our memories to persist and to be accurate, we also have some that we wish we did not have. Wouldn't it be great to be able to forget some things willingly? To just clean things out and erase some memories? Our memories often have the unfortunate characteristic of being persistent when we don't want them to be. Traumatic events, unhappy events, and other events that we would like to forget are often difficult to forget because of their initial salience or emotional content, or because of intrusive recollection and rumination. Every time we think about that unfortunate event, we end up strengthening the initial trace. Dwelling on an unfortunate event in the past is likely to make that memory trace all the stronger. Try not to think about something and you end up thinking about that thing even more.

These seven sins are fairly broadly defined. All of them are examples of memory errors that come about because of how our memories are structured. All of these come about because of how our memories work. Our memories guide us, but they also sometimes mislead. These can be minor annoyances, but they are all around. Try to recognise them in your own daily interactions. You won't be able to avoid these seven sins, but you can learn to recognise them. And if you can recognise them, you may be able to help to keep them from interfering with things.

For example, have you ever been shopping at a farmers' market or outdoor city market? Think about this scenario. Imagine that you have entered the market, you have examined the fruits and vegetables available at one of the stands, and then you have moved on to another stand. You're shopping with a friend and you each have a few things that you would like to buy. As you move through the market, and to another stand, your friend asks if the first vendor had lemons for sale. *Did they have lemons? Now that's a good question . . . Did you actually see lemons or do you just think you saw lemons?*

Questions like these are often very difficult to answer because you take so much for granted when you are perceiving things that you might not commit the common details to memory. You end up filling in a lot of detail from memory in general that may or may not have been present during the actual perception. As a result, it can be infuriatingly difficult to answer a question about what you just saw. The stall may have had lemons, if they had other fruits. But you may find yourself not being 100 per cent certain that you actually saw lemons there. Perhaps you are just inferring the presence of those lemons based on your knowledge of produce vendors in general. In other words, your memory representation of a produce vendor may have a lemon feature. That feature will be activated when you visit the produce stand, even if you don't explicitly see the lemons. Your memory, in this case, might be making a mistake even as it tries to help by filling in details and saving you time and energy.

We are, for better or worse, using our memories to perceive, to guide behaviour, to make judgements and to assess situations. We cannot live in the present without being influenced by the past. Given the primary importance of memory in everything we do, I think it's time we dig deeper into how memory works from a cognitive psychological perspective. In the next chapter (Chapter 7) I'm going to take a look at the idea of memory systems and the evidence that we have different kinds of memory and then take a closer look at a theory of short-term memory called working memory. Then in Chapter 8, I'll take a closer look at how we organise our long-term memory into memory for facts and concepts and memory for personal events.

CHAPTER 7

Working Memory: A System for Thinking

In Chapter 6, I spent a lot of time on all the ways that our memory seems to lead us astray, including memory errors, heuristics and biases. Memory is one of the most important things we have in our cognitive toolbox, but it's not very reliable. However, I did not write much about how psychologists study memory, how memory works and the different kinds of memory systems that we have. That's what these next two chapters are about. If you want to make good decisions, accurate judgements and useful predictions about the world, you need to be able to use your memory effectively. If you want to avoid the pitfalls and errors that I discussed in Chapter 6, you need to be able to use your memory adaptively. Good, adaptive memory use requires a good understanding of how memory actually works. So in this chapter, we'll focus on a common distinction in memory: the difference between short-term and long-term memory, and then I will describe and explain a particular theory of short-term memory called the 'working memory' model that shows how short-term memory is closely connected to perception and is moderated by attention. I'll have more to write about long-term memory in Chapter 8.

Most psychologists (and most of us) think about memory as a set of interdependent systems. We often fall back on a computer analogy for memory because we recognise that a computer has both active memory (RAM) and also long-term storage in the form of the hard drive. This idea allows for a close, analogic relationship with our intuitive sense of short-term memory and long-term recollection. Just as in the computer, the former stores information you are currently working on and the latter stores files that you might need later. This computer memory and file storage metaphor are useful for thinking about these broad generalisations, but it's not really the right way to describe our memory, because at a fundamental level for us, all memory is just memory. All memories are reconstructions of events and experiences that happened before. But we use memory in different ways. We access our memories under different conditions. These functional differences lead to what might seem like systematic differences, and so we can discuss memory at that level.

We often use other metaphors beyond the common computer metaphor for memory that suggest we think about memory as file storage in a more general sense. We say things like 'file that away for future use', 'commit this to memory', or even 'seared into my memory' which really suggests an indelible imprint. But memory is not really like that either. Memory is more dynamic and active than a file on a hard drive and it's more dynamic than a single event searing or imprinting itself. As we saw in Chapter 6, memories are constantly changing but, despite that constant change and fluid nature, we use memory for everything we do: from keeping track of the moment-to-moment activity in front of us and remembering facts about the world, to remembering what happened yesterday and making plans for the future. Because there are so many different functions for memory, psychologists have studied it in many different and dynamic ways. The end result is a very large and complex scientific literature on what memory is. We can make sense of this complex study in several ways. One way is to describe the core functions that seem to be common to different memory systems and theories. Another way to try to make sense of memory literature is to look separately at some of these differences and divisions. That is, consider all of the ways that memory can be described, divided, conceptualised and studied. By looking at memory in these two ways, we will understand some of the common principles (how memory works in general) and some of the specific principles (why there seem to be different memory systems).

Basic Memory Functions

Your memory has several basic functions, but its primary function is to allow you to do more than just react to what is directly in front of you. Memory is what allows you to learn things and to generalise from past experiences. But how exactly does our memory allow us to do this? Let's define three basic operations that your memory carries out: encoding, storage and retrieval. Encoding is the process of putting things into your memory. To encode something implies that the brain needs to alter the format of what you are perceiving and to put it into a different code. This encoding process is a reconstruction of perception. The nature of this encoding process implies that your memory often has a strong connection to perception. The most effective way to encode something, then, is to try to reactivate as much of the original perceptual experience as possible. To some degree, that's what your memory does: you perceive something, which activates your brain in a certain way. By encoding, you will attempt to maintain that activation long enough to be able to store and retrieve the activation for later use.

A second function or basic operation is storage. We use our memories to store this perceived and encoded information. It's not a storage system like a wardrobe or a file system on a computer which would imply that there is a physical place for each memory. Rather, your memories are stored as connections between neurons and the information is distributed across different areas of your brain. We need to store things for different amounts of time. Some information is only stored for a few seconds (or Op cut if necessary to fit 'wardrobe' above less); other times, we store and reactivate the same memories for years and decades.

The third function is retrieval. Retrieval is using your memory. It can be in the form of explicit recall ('I remember this fact') or an implicit retrieval in which a previous experience influences how you behave towards something in the future. The retrieval can be in the form of filling in the details of a perceptual scene (as we discussed in Chapter 6). Or retrieval can take the form of mental time travel. We can experience a vivid recollection of a previous event that happened to us, like getting your first job, a first date, the birth of a child, or even a mundane event like the last trip you made to the shops.

These three functions, encoding, storage and retrieval, describe what our memory does. But there are many different circumstances that affect these functions. The end result is that we seem to have different memory systems, some of which are more closely tied to encoding and others that are more closely tied to how things are stored and retrieved. Although all of our memory is in one place (our brain), we seem to have different kinds of memory and different systems of memory.

Different Kinds of Memory

When I write, I usually sit at my home office desk. Sometimes, when the house is empty or if I am awake early in the morning, I might sit at the kitchen table. In the kitchen, there is a sliding door and window leading out to the deck where we have a small garden, a pond and some bird feeders. I'm not an expert in gardening. I'm not an expert in birds, either. But I do like to see all the action at the bird feeder in the summer. If I look out at the feeder and notice a bird, my memory is instantly and automatically involved in the process of looking, recognising and thinking about that bird. Let's consider all the ways that memory is going to be involved in this deceptively simple action and then let's organise our discussion of memory around this.

To begin with, even the basic action of mindlessly turning to look at the bird feeder involves some form of memory: I already know where to look and what I am looking at. This behaviour is automatic and motoric: my head turns almost without noticing and my eyes look where I expect the feeder to be. I don't need to involve consciousness to do this. I don't need to be aware of the memory needed to turn and look. This is a *procedural memory* and it's the kind of memory that we rely on when we remember the actions needed to ride a bicycle, remember how to type or pick up the coffee cup without thinking. And this procedural memory guides my eyes away from my laptop to the birdfeeder behind the house and I focus on the birds at the feeder.

I see two small birds sitting on the peg of the larger feeder. Without even giving it much thought, I already know they are birds, because the perceptual input will activate a network of neurons that I recognise as being the same network that is active anytime I see birds. That, too, is a function of memory which we'll call *sensory memory*. This sensory memory, as I will explain later, lasts only a fraction of a second and it's really just enough to keep the perceptual input active until other cognition happens. If I want to do anything else with this visual input, I need to keep this representation active long enough to recognise what I am looking at. I need to keep it active if I want to think about the birds that I see. Keeping a short, sensory input active involves an active form of memory known as *working memory*. Working memory is what we usually call short-term memory. Working memory is what connects perception with our active processing of what is in front of us. Working memory is the memory that holds what we're actively working on in consciousness.

Let's recap. I use a form of automatic and procedural memory to look at the bird feeder. I focus on some birds and activate a sensory memory system and I keep that representation active by devoting some of my cognitive processing resources in the form of working memory. So far, so good. And so far, this is all happening automatically, without any intention or will. That comes next.

Now suppose I am actively thinking about the birds. And suppose I really get caught up in watching these birds (easy to do . . . birds are fascinating to watch). A few things might happen next. First, I might keep the working memory active by thinking about these birds and maybe trying to identify them. The effort required to do this might mean that I withdraw from what I was doing before I noticed the interesting birds at the feeder. [25] I might even forget what I was writing about and lose my train of thought because I'm now using my working memory to think about these birds. The contents of working memory can be maintained, but working memory is limited to a small window of active experience with the world. We can only think about a few things at once. When I switch to thinking about the birds, the result is that I clear everything else out of my working memory. I have now forgotten the sentence I was writing and I'm thinking about the birds at the feeder in my garden.

The first step in this active memory process is that I focus on the birds. The second thing that happens is that the representation of those birds in my working memory, which includes the visual input and the image of the birds and maybe the sounds they are making, is going to make a connection with memories that have been encoded and stored. What I perceive and what I am maintaining in my working memory (a representation of a bird)

---

25 Does the phrase 'withdraw from what I was doing' remind you of anything? It might remind you of the quote by James on attention in Chapter 5. If it did remind you, take a minute to think about that reminding. Was there an explicit recollection of reading it before? Do you remember reading it earlier in this book and remember where it came from? Or was the reminding more of a vague feeling of knowing that it sort of reminded you of *something* that you had seen but did not activate a specific memory until I mentioned it just now. If either of these things happened, it is a good example of explicit and implicit retrieval. It's also a good reminder of how memory and attention are closely linked. There is always some cost to switching and withdrawing. In fact, if you took the time to read this long footnote and think about whether or not you remembered that James quote, you might have even forgotten that I was writing about looking at birds and how I get distracted from my own writing when I withdraw to look at the birds at the feeder.

will connect with my concept of birds. This process activates a kind of memory called *semantic memory* which is my memory for facts and things. Semantic memory holds all the things I know in an interconnected web of concepts and ideas. A representation that's active in working memory will probably find some matches with other representations that have been previously active. These representations are all connected with each other so that this unique experience that we are having in the present (watching this bird right now) will overlap with representations for knowledge acquired in the past. Some of these activated memories will be closer to each other as a function of perceptual and conceptual similarity. These activated memories will also connect with the name of what I'm looking at ('bird') and maybe even the kind of bird ('black capped chickadee'). The general and specific knowledge is one way that a semantic system can be organised. These general and specific representations connect with a form of memory known as *lexical memory* which is the part of our semantic memory that stores words that are connected to these concepts.

Semantic memory, lexical memory and concepts are all part of a larger kind of memory that is sometimes called *declarative memory*. This is a memory that you can 'declare' the existence of. Unlike the more automatic forms of memory like procedural memory or sensory memory, which do much of their cognitive work outside the realm of consciousness, declarative memory is memory you can probe and investigate. You can think about what's in there. And as we'll see later in the next chapter, what's in there is a dense, interconnected network of facts, ideas, places, things, names, words, images, concepts, sounds and features.

Let's return to my example of watching birds at the birdfeeder. Suppose these are especially interesting birds. Suppose these are new birds I have never seen. Or suppose they are birds that seem familiar, but I don't know the name. Making this determination involves memory in a few more ways. First, I need to recognise that I do not recognise them. I need to recognise the limits of my knowledge. This is a form of higher-order memory awareness called *metamemory*. Metamemory is an awareness of what I know and what I do not know. We use metamemory to make decisions about whether or not we are aware of what we know, and we use it to make decisions about whether or not to commit more processing to the behaviour and process of figuring out what birds these are.

In addition to metamemory guiding my decisions about whether or not I know the birds I am looking at, there may also be other thoughts and behaviours that rely on memory. Let's assume that at this point, I've completely abandoned the work that I was doing. I've looked away from my computer and I am focusing on the birds. I've forgotten what I was writing about and I'm now totally committed to watching these birds. As I search my semantic and lexical memory for the name of the birds, I may need to search my memory for images or representations of them. This is very likely going to trigger other connections beyond the general and specific conceptual knowledge that's in my semantic memory. Perhaps I recall some other time and place where I saw these before. That might trigger a really specific memory for seeing a bird like this on vacation or at the feeder last year. This activates yet another kind of memory referred to as *episodic memory*.

Episodic memory is that kind of memory that we probably have the strangest relationship with, because unlike our semantic memory or even the procedural memory, where broad generalisations are tolerable and even helpful, we usually want our episodic memory to operate with great accuracy. It does not. I might now be trying to remember if I saw these birds on vacation last year, two years ago, or at some other place all together. Are these memories even accurate? The discussion from Chapter 6 suggested that they are not always accurate, even if they are useful.

Let's recap yet again: procedural memory, sensory memory and working memory underlie the behaviours of looking, registering and maintaining the perceptual experience of looking at the birds at the feeder. I maintain the image and thoughts in my working memory. This working memory connects with my semantic memory to recognise the birds. My sense of metamemory tells me that I do not immediately know what kind of birds these are. As I search my semantic memory, it activates and stimulates recollections of past episodes, some accurate, some not (again metamemory may help disambiguate). But that is still only part of what might happen. All I've done so far is spend a few minutes losing track of what I was writing and looking at some birds, getting caught in a daydream. I can use my memory to be more productive even in this moment of lost attention.

Perhaps I might now think about needing to fill the bird feeder at some time in the future. Do I have enough bird seed? Should I remind myself to order more? When will that run out? In this case, I am using my memory to think about the future. I am using memory to plan for and to guide future behaviour. This future-oriented form of memory is another form of episodic memory known as a *prospective memory*. This kind of memory is a memory for things in the past, used in the present, to make plans and predictions for the future.

All of this depends on our mental and neural systems being able to encode, store, retrieve and use information. In the next section, I want to go into more details about how that works. Let's begin with a structure found in the temporal lobe: the hippocampus.

The Hippocampus

We know that memories are stored in your brain. We know that a memory is not stored like a file in a drawer and that the term 'storage' is a metaphor. In this case, the storage is distributed throughout the brain. This distributed storage takes the form of connections among the network of neurons that were active during perception. These connections are strengthened as the memory trace is strengthened, which allows them to be reactivated more easily. Strong connections correspond to memories that are able to be retrieved quickly and are retrieved often. This quick retrieval helps you remember things from the past when you want to relive them. This quick retrieval is also what is responsible for memory in filling in details and helping you to complete a perceptual experience. Frequent events are encoded strongly by reinforcing connections between neurons. These strong connections also result in memories that are able to be available to influence behaviour and these underlie the availability heuristic that I wrote about earlier in Chapter 6.

This basic process of activating neural connections, strengthening the connections that occur frequently, and being able to retrieve information for later use depends on a structure I discussed in Chapter 3: the hippocampus. The hippocampus is located in the subcortical regions of the temporal lobes. The hippocampus is part of a system that connects perceptual input from the sensory organs, with attention and with memory. The hippocampus helps you store new memories and use your memories to interact with the world. It does this by pulling together information about what is happening in the brain and where it's happening. It is then able to recode this information so that it can be reactivated later.

Exactly how the hippocampus does this is definitely still a matter of scientific debate, but one theory being investigated by Joel Voss and Neal Cohen (Voss, Bridge, Cohen & Walker, 2017) is that the hippocampus connects to and receives connections from the amygdala as well as brain areas that encode *location*. They note that several studies have found very strong connections between guided eye movements and activity in the hippocampus. The hippocampus is not directly involved in coordinating eye movements at the level of motor control, but rather, it seems to be involved in how we use memory to know what we are looking at and where to look. The hippocampus seems to be able to activate and reactivate connections in the brain that connect knowledge with perception (and vice versa). It seems to be especially attuned to where you are looking. This research seems to support the ideas that we are constantly using memory, via the hippocampus, when we're perceiving the world.

Voss and Cohen also discuss research that corroborates this visual, 'exploration and attention' role for the hippocampus. They note that research relying on brain imaging techniques like fMRI finds correlations between activity in the hippocampus and experimental tasks involving memory-guided looking (e.g. tasks that require the participant to look at some place they looked before). It seems that the hippocampus is critical for helping us guide our perception. According to Voss and Cohen, it does this by creating brief-interval memory signals that we then use to guide our visions. The hippocampus creates 'online memory representations' that are broadly consistent with the ideas I've been discussing in this chapter and earlier in Chapter 6 about the role of memory in filling in the details of perception.

This research from neuroscience shows that the hippocampus is critical for this blending of perception, attention and memory. Without a fully functioning hippocampus, we would have difficulty knowing where to look, knowing what we are looking at, and we would not be easily able to use memory to fill in these details. That's not everything the hippocampus does, but it's a critical piece of the puzzle of how we get from a constantly changing perceptual input to a stable understanding of the world and a stable memory representation.

This research provides a glimpse into one of the functions of the hippocampus. But this glimpse raises more questions. For example, what is the form or nature of this memory representation that the hippocampus creates? Are there different kinds of memory representations? Voss and Cohen are suggesting that that the hippocampus creates and manipulates short, active and online [26] memory representations. But they also assume that these short active memory representations share some correspondence with other representations in perception and in long-term memory. The hippocampus is creating these representations to bridge perception and long-term memories.

The role of the hippocampus is still being studied, but we now have a more complete picture of how memory works to help perception and how it guides action. I've laid out a general organising framework for the role of memory in perception, active memory and long-term memory. These three kinds of memory have long anchored our colloquial understanding of memory and thought: We perceive, we think, we remember. We understand memory as a sensory phenomenon, a short-term phenomenon and a long-term phenomenon. But at the same time, we also need to consider memory as a function of its contents. Memory for specific events in our past seems to have a different character than memory for well-rehearsed facts such as the memory we use to recall and apply the correct motions to unlock our iPhones.

I'd like to spend most of the rest of this chapter on some of these divisions and look at how memory representations work. We can do this in a number of ways, but I think it's best to start with divisions of memory by duration (short, medium and long) and content (events, facts and actions).

Memory Duration

At a basic level, we have memories that last only a short time – a few seconds – and other memories that last a lifetime. Can you remember your home

---

26 You may have noticed that I have been using the term 'online' here. The term 'online' is often used to refer to representations that are changing, dynamic and closely connected to perception. Like so many aspects of cognitive science, this term comes from the computer metaphor for the mind. An online representation is one that's dynamic, created as it is needed, and kept active. The metaphor refers to a representation that is like being online and connected to the internet instead of accessing static, stable files that are on a hard drive. When you're online, you're live. When you're online, you're sensitive to all the dynamic, changing information on the internet. This term is also often used to refer to methodological techniques like fMRI, eye tracking and EEG, because the measurement is active and dynamic and not filtered through the lens of self-report. Self-report measures and other kinds of behavioural responses are 'offline' because the response is made after the stimulus is presented, whereas brain imaging can capture activation as it is happening online. The computer metaphor permeates cognitive science because it's part of the creation of cognitive psychology and cognitive science.

telephone number from when you were a child? There may come a time when this question no longer makes sense to people, but many of us grew up with a 'landline' phone that we might think of as our home phone. I still remember my parents' phone number. It seems to be a part of who I am. It is almost like my name. Almost, but not quite. Unlike my name, which I need, I can still recall my parents' phone number even though I no longer need to remember it. I have not called it in years. I'm not even sure it's still an active number to be honest, because when I call my dad to talk, I call his mobile phone by looking up the contact information on my phone. He's had that number for years, but I need to look it up each time. In other words, I remember the old landline phone number from the 1970s and 1980s even though I no longer need it. And I do not remember the newer wireless phone number even though I use it.

But there must have been a time when I didn't know that old number and I had to try to learn it. I probably did what most of us do, I repeated it to myself using my inner voice until I could remember it. If I ask you to remember a phone number, let's say '958-8171',[27] how would you do this? You probably say it to yourself. You could try that for yourself now, try to remember that string of numbers without doing anything else. Read it to yourself, close your eyes and see if you can remember it for a minute accurately.

If you're like most people, you closed your eyes and you said the numbers to yourself, and as long as you kept saying them to yourself, you'd be able to remember the string of numbers. If somebody came and interrupted you, you would probably forget almost all of them. This is the experience that most of us have with what we call short-term memory, it is short, it has limited capacity, and we are most often aware of it when we are rehearsing things with an inner voice.

What do you do when someone asks you to remember a list of information? Lists of information are kind of an interesting data structure. A list is something that is usually in an order, it's usually short, and it has a very specific intention. For example, a shopping list. Everyone knows what a shopping list is. You write down the things you need to buy, you take the list with you to the shop, and you cross off the things as you go. You can do this without a written list of course, someone could ask you to get milk, eggs,

---

27 This is not my parents' old home phone number, by the way, I chose these numbers at random using the useful and fascinating website http://www.random.org, which uses atmospheric noise to generate true random numbers. If this is your phone number that's just a purely random coincidence.

bread, spinach and Parmesan cheese. It shouldn't be hard to remember that list. How would you do it? As with the phone number you would probably say the list to yourself a few times, and then drive to the shop, and then try to retrieve that list from memory. Retrieval is a metaphor. The list is not actually being stored in a place in your brain that you go to and pull the memory out of. But rather, you are re-creating that particular event or experience. You are probably going to remember the list items in the same order. In fact, without looking back at it now, see if you can remember the list in that order. You probably will.

Our memory likes lists and it likes order. This is one of the reasons we make lists. We use the structure of the list itself as a memory cue. Each item is closely associated with the items that preceded it and the items that come after it. The list is its own form of reinforcement. If you want to remember a short listing of things, try to keep them in a specific order.

So different memories have different durations, depending on what you use them for. Your parents' phone number might stay with you for years. Lists are stored as long as you need them (and can be kept active). Other memories are even shorter but critical in helping you hold a stable perceptual experience in your mind. We call these sensory memories, and this is where your memory meets the world and where the world meets your memory.

Sensory Memory

Short-term memory seems to come in several different varieties. At the lowest, most perceptual level is what psychologists call 'sensory memory'. This reflects the activation of brain areas most closely related to those that are active during the act of perception. This was discovered by the mid-twentieth century psychologist George Sperling in what was one of the most creative and clever experiments of its day. This preceded the modern cognitive neuroscience research by decades and helped to provide an empirical demonstration of how we use our memory to fill in details in the world and to guide perception. Sperling's clever experiment worked in the following way.

Research subjects would be shown an array of letters on a screen for a very brief time, less than a second. For example, suppose you are asked to view the letters below for about 50 milliseconds, just a tiny fraction of a second.

The letters would flash and then disappear and be replaced with a block of visual noise to make sure that the image would not linger. This can be controlled precisely with a device called a *tachistoscope*. The tachistoscope uses slide projection and a mechanical timer to present images for an exact amount of time. The same thing can be done on computer monitors now, but not in Sperling's time.

G	K	Y
W	P	B
R	T	H

When the exposure to the letters is very brief (the 50 msec typically used), subjects could never recall all of the letters. They could get three or four and then the image seemed to fade. But the subjects indicated that they had the sense and the sensation, that for a very brief time, that all of the letters were visible but that they faded from memory before they could be recognised. I often do a demonstration of this experiment when I teach about memory in my university course. I present the array on the screen for a split second and then advance the slide. No one can remember more than three letters. Many students report the same experience that Sperling's subjects did. They say they can see the letters all at once, taking up space in a visual mental image, and then they have to read them to remember them. But as they try to read the letters from their visual mental image, the image fades and they can only get to three letters.

In other words, people can see all the letters, and they have a sense that they were all represented in some way at the earliest perceptual level, but they can't seem to stabilise the perceptual experience before it fades. There is a full image available for a split second, but there is not enough time to fill in the details. That's what seems to be happening to subjects in the experiment, but how would you be able to know that for sure? Sperling's solution to that question was very clever and, in many ways, changed our understanding of how to study perception and memory. Rather than rely on his subjects' introspection of their mental state (the sense that they could see all the letters before they faded), he devised a way to measure this phenomenon. In the standard condition, which Sperling referred to as the *full report*, subjects would try (and fail) to recall all the letters. But in the experimental condition, which Sperling referred to as the *partial report*, they did not have to recall all the letters.

In partial report, subjects saw the array as before and immediately after it disappeared, they heard a tone that was either high, medium or low in pitch. When they heard the high tone, they were to report only the top row of letters. When they heard the medium tone, they were to report the middle row. When they heard the low tone, they were to report the bottom row. Notice that the tone is a different modality, so it does not compete for the same neural resources as the visual image. Notice too that the only way in which you can report a row is after it disappears and after you hear the tone. You do not know which row to report until after it disappears from the screen. The only way a person can do this task is if they really did have the full image represented in sensory memory and that they could inspect the memory image as directed by the tone.

Sperling's subjects were able to report back all of the letters in the row that was cued by the tone. And because they could do this across several different trials, for all the tones and for all the rows, it suggested that they did indeed have all of the visual information needed to perceive all of the letters for a very brief time, but that it was not long enough to see them all. When I do this example in my course, I see exactly the same thing. People can report back any row, even though the image disappears from view before the tone is played.

There are so many good points in this experiment, that I think I could teach a whole class on just this. First, this experiment is another example of what I discussed in Chapter 5 on attention. At the lowest level, there seems to be an unlimited capacity in our perceptual system. This experiment also makes a point that visual and auditory attention do not tend to compete for the same resources. These are, for all purposes, two different streams. You can see the image and hear the tone and those two sources of information do not interfere. This experiment also illustrates just how much our perception depends on our memory to fill in the details. In this case, the subjects literally cannot perceive a grid of letters until they have activated a memory representation for each letter and in effect read the letter to themselves. There is unlimited capacity low in the system but that unlimited capacity does not mean very much if there are no concepts or memory to activate.

This is one of the most challenging paradoxes in psychology: We can only perceive what we know, and we can only know what we perceive if we know it already.

The sensory memory of the kind that was discovered by George Sperling is only one kind of the short-term memory. It lasts only for a split second. That is, unless we decide to do something with it, the information fades pretty quickly, in a few hundred milliseconds. Sperling's experiments confirmed that as well. But once we read the letters, words, numbers, or images, we can remember the information for a longer time and can maintain the letters in our conscious awareness. This is the kind of short-term memory most of us are familiar with. It's the kind of memory we are used to working with when we think, solve problems, and have to keep several things straight. The memory that does a lot of work for us is called working memory.

Working Memory

While Sperling's sensory memory is a brief flash, the kind of short-term memory that we call 'working memory' seems to involve work. Working memory is a form of active memory that is short in duration and/or short in capacity, and it reflects the information that we are actively working on or actively thinking about. Working memory is the memory and thinking that we are consciously aware of, although we may not always be consciously aware of everything that is in our working memory.

Let's look at some examples before looking at the details. Suppose you are listening to a podcast or a lecture. You're paying attention and following the story, but it's still new information. You can make assumptions and predictions about the gist but don't know exactly which words or sentences will come next. However, you need to hear and understand each word so that you can understand the sentence and the idea before the initial perceptual representation fades. And it will fade quickly. Sound is only around to hear for an instant. As soon as you hear it, it's gone. As soon as you perceive it, it's unavailable. There is an awkward problem to solve. You hear words that fade as soon as you hear them, but you also need some time to fit these words into a sentence and into the gist of the narrative. You also need to solve this problem very quickly, because you are hearing new words all the time. You have to work to keep up! This is the same problem that

Sperling was studying, by the way, except he was looking at the existence of brief visual sensory memory.

If only you had a short-term 'holding area' that let you keep the perceptual form of the word sounds active for just a few seconds so that you could fit them into the sentence, phrase, or concept that you are listening about. That would solve the problem of the rapidly fading signal that is perpetually in danger of being overwritten by new words that you hear before you can comprehend the ones you just heard.

Well you do have such a system. That's essentially what your working memory system is and what it seems to have evolved to do. It's a short-term holding area for information that is closely tied to perception but also conscious and active and able to act as the intermediary between perception and knowledge. We rely on this system for much of our active cognition and thinking.

Our working memory handles active memory of all kinds. Let's look at a few more examples. Consider the mental activity that is needed to read this passage of this book. As you read each word and phrase, you activate mental representations that help you extract meaning and build some kind of mental model for what you are reading. You use your working memory to do this. Your working memory stores the information initially, before you fully know what it means, so that you can quickly activate and access the concepts you need to help you build a mental model. Now I realise that reading, being a visual process, does not face the same problem that listening does. Written words stay right there on the page, unlike the spoken words that disappear as soon as they are heard. Although reading is a visual process for most of us, it still activates the parts of our brain that are active for spoken language. When you read, the words still pass through the working memory system. You read with an inner voice that helps you hold on to just enough of the signal (the visual input or the sound) so that you can begin to activate concepts and ideas.

Working memory is not just language based. If you are watching a video, looking at a picture, or watching a scene in front of you, the images you perceive can be actively maintained in your working memory before you connect them with other concepts and form an idea. When you solve a problem, such as a mathematical or physics problem, you might notice that you need to keep several ideas simultaneously active until you can put them together to solve the whole problem. You might talk to yourself, with an inner voice, but you might also imagine a three-dimensional object and then imagine how it moves or how it looks from a different angle. Can you solve a Rubik's cube in your imagination? That's working memory. Can you keep track of the location of several footballers on the field without having to list everyone all the time? That's working memory too. Can you perceive and identify visual objects in different locations? Still working memory.

It should be clear that working memory plays a big role in our understanding of the world and in how we reconstruct experiences to perceive and comprehend what is going on around us. Let's discuss how this model works in more detail, the theory behind the model, and some of the evidence that supports it.

We've already discussed some aspects of this theory in a previous chapter without using the term 'working memory'. For example, the rote rehearsal that you might engage in to learn a phone number or a list of words uses an inner voice that is an integral part of the working memory system. In one example from Chapter 5, I described an experiment that was designed by Lee Brooks, and I mentioned that doing this task required the use of two 'attentional pools'. Subjects were asked to maintain a visual image or repeat a sentence, and they were further asked to answer questions by saying an answer or pointing. Brooks observed interference when the memory used the same 'attentional pool' as the response. Brooks' work showed that there is both an 'inner voice' and an 'inner eye' that most people can use, actively to maintain representations. Brooks' work did not refer to this as working memory, but his work paved the way for the later development and refinement of the theory.

The working memory model that we'll discuss is a very specific theory of short-term memory that was originally developed by Alan Baddeley and Graham Hitch in the UK in the early 1970s (Baddeley & Hitch, 1974). Although there have been a number of theories of short-term memory that are similar, the working memory model is probably the most well developed. There are a few key aspects to this system. Baddeley's working memory model assumes that there is a system or network of neurological structures that process immediate sensory information. The working memory system acts as a buffer so that the information can be maintained, processed further or discarded. Baddeley's model has several features and assumptions that are unique to this theory. These assumptions explain our behaviour in a number of different scenarios.

First, and in line with what we've been discussing about the close connection between perception, attention and memory, Baddeley's working memory model is modality specific. In this case, 'modality' refers to the perceptual modality. This is just a way of saying that there are different memory systems for different sensory systems. The working memory system assumes that there are separate neural circuits for the active processing of auditory information and separate neural circuits for the active processing of visuospatial information. Auditory and verbal information is handled by a system that Baddeley and Hitch called the 'phonological loop'. Visual and spatial information is handled by a system that they called the 'visuospatial sketchpad'. The coordination is handled by a 'central executive'. The central executive allocates resources, switches between systems, and coordinates switching between and among representations within a system. Baddeley's model is not the only model of working memory, but it is one of the most well-known, and the other theories of working memory make some of the same the assumptions.

According to Baddeley, the phonological loop is a phonological or acoustic store that is connected to input from the auditory cortex. As we saw earlier with the description of sensory memory, an initial memory trace will fade after about two seconds. But we don't usually experience the world like this. We are able to focus on one thing we hear and keep thinking about it. The things we hear don't seem to fade, because we are able to maintain them or reinforce them by way of an articulatory control process known as subvocal rehearsal, that is, the inner voice.

The inner voice is familiar to most of us. Think about how you can remember a short piece of information (a number or a short phrase) while you repeat it to yourself. For example, if you use two-factor authentication for a website or a social media account, and you log in from a new machine, the website will text you a five- to eight-digit number that you enter in the website. You might try to remember the string of numbers by repeating them silently to yourself until you type them into the box, and then you immediately forget them. The subvocal rehearsal maintains them in your memory as long as you keep repeating them. They are forgotten as soon as this rehearsal ceases and the memory fades.

The inner voice is also part of how we listen and comprehend spoken language. According to Baddeley's model, everything we hear passes through working memory on its way to being understood later. We're not usually aware of working memory in this context, but sometimes we do notice its influence. Has this ever happened to you: someone asks you a question or says something to you, but you do not quite hear what they say? When this happens, you might ask them to repeat the question. You ask, 'Sorry, what did you say?' But sometimes, before they can even answer you, you end up sort of 'hearing' their question anyway by replaying what they said in your mind. You are able to hear what they said, a few seconds after they say it, by using the acoustical control process of the working memory system. In this example, you are replaying the message that you heard *before* you know what was actually said. This implies that the verbal working memory system is both language (phonology) based and acoustically based. We are able to replay sounds in our minds, not just words. Working memory allows you to reactivate the auditory cortex to rehear what was said.

This inner voice (verbal working memory) is a crucially important component of thinking. We need an inner voice and verbal working memory to formulate hypotheses, to read through a problem description, to frame a decision, to test alternatives, and to consider the outcome of our behaviours and actions. In other words, active thinking needs an active working memory. Active thinking requires language comprehension, and working memory for reasoning, planning and problem-solving. In fact, success in all these activities has been shown to be highly correlated with measures of working memory capacity (Oberauer, 2009). In fact, for many psychologists, 'working memory capacity' has become kind of a catch-all feature of intellectual capacity, and later when I discuss some of the so-called 'executive functions', we'll see why.

The evidence for the existence of a phonologically based, verbal working memory system is extensive and diverse but one of the most striking demonstrations of how the whole system works in conjunction with long-term memory is the serial position effect. The serial position effect, which you probably already know about, is one of the foundational effects in cognitive psychology. This effect was discovered in the very early twentieth century by Hermann Ebbinghaus. Ebbinghaus must have had an easy lab to manage. He mostly tested himself and he spent a lot of time memorising lists of words, letters, strings of letters under different conditions and he carefully recorded his performance on recognition and recall. The serial position effect that he discovered is deceptively simple. I can do this as a class demonstration and it will replicate every time. It works like this. The participant in the experiment is presented with a list of words (or syllables, numbers or letters, etc). This list is long enough that they will not be able to remember all of them, but not so long that it will take more than a minute or two to read or hear them. This would be around twenty words, presented at a steady rate, about one per second. If you were a participant in an experiment like this, you would be asked to remember as many of the words as possible. And you would try to remember the list the same way that you try to remember phone numbers, a shopping list, or any short list: you would use your inner voice (your phonological working memory system) to repeat the words to yourself to try to maintain them.

There is a problem, however. The list is too long to be able to fit in one 'loop' of the phonological loop. There are just too many words to say and at some point, if you try to fit too many words in the loop, you will forget some of them. And there is another problem. As you try to say the words to yourself, you are also being presented with additional words on the list. These words will keep coming at you while you are trying to repeat the words that you already heard or saw. You will have to decide either to keep repeating the set of words you already have or stop trying to remember those words and try to remember the new ones you are being presented with. One way or another, your cognitive system can't handle all of this information processing at the same time. It can't handle all the words, at a steady pace, while trying to repeat them all. Some of the words will not be learned. But the interesting thing is that there are some predictable ways that you would fail to learn all the words. The pattern of errors is related to the way the working memory system operates.

If you are presented with this list of words (auditorily or visually, it works either way), you will not be able to remember the whole list. But you will be much better able to recall words from the beginning of the list and/or the end of the list. That is, your memory will be sensitive to the *serial position* of the word in the list. The order and context matter. When Ebbinghaus first carried out this research, he found that he was able to remember the words at the beginning and the end much better than the words in the middle of the list. Hence the name 'serial position' effect. There is an effect of the serial position of the words on the list. Ebbinghaus did not describe this as a product of a working memory system, but his explanations and the more refined explanation provided by the Baddeley working memory model are roughly the same.

As you repeat the words at the beginning of the list to yourself in your working memory, they form stronger traces. They will seem to be more strongly committed to your long-term memory because you keep repeating them even as you hear and try to repeat subsequent words. The repetition helps to develop stronger traces. As long as you are able to keep rehearsing these words, you can keep them in your working memory long enough to recall them during the test phase later. Better memory on the items early in the list, usually called the 'primacy effect', is an effect of working memory rehearsal and the stronger traces that result from the repetition. However, you can't rehearse the middle items as well, in part because your working memory is already full, from rehearsing the early items. Your ability to remember those words is impaired. The primacy effect is limited by the span of your working memory.

Ebbinghaus found that he was also able to remember the last few words on the list. This is called a 'recency effect' because the memory enhancement is for the most recent words. Unlike the primacy effect, which is a result of the extra memory rehearsal, the recency effect might come about because these words are still in the active, acoustic working memory store. The working memory model is able to account for the primacy and the recency effects by assuming that the former is the result of rehearsal and the latter is the result of the word forms still being active from the sensory input. The words from the middle have not been rehearsed because there was no additional capacity in the phonological loop to rehearse them. And they are no longer active in the sensory store because they keep getting pushed out by each new word.

This explanation works well enough for the primacy and recency effects in a standard serial position effect. But there are two variations of this experiment that provide even more support. Variation 1: If the words are presented at a faster rate, two words per second for example, the primacy effect will be reduced or eliminated, performance on words from the middle of the series will also be depressed, but the recency effect will not be. Why?

The faster rate will overwhelm the phonological loop early on. You would have less time to rehearse the words. It takes time to rehearse words in your phonological loop, which tends to operate at normal speaking rate. Variation 2: If the presentation of the list is followed by a 20 second delay that is filled with additional words, there is no effect on the primacy effect, but the recency effect is eliminated. This is because those words at the end of the list are pushed out of the active memory by the additional words in the delay. A 20 second delay without the extra words will not have as strong an effect, because some of the final words will be able to be rehearsed.

These serial position effects may seem artificial and circumscribed. They may seem like effects that can be generated in a laboratory but might not generalise outside the lab. But if you stop and reflect on some examples from everyday experience, it's a reliable effect outside the lab as well. For example, have you ever heard advertisements on the radio that end with a very fast spoken bit in 'terms and conditions?' It goes by so fast that it's impossible to process through your working memory and so you rarely understand everything that was said. This is just like the fast list presentation. There is not enough time to encode and process all the words, so some of them never make it past the initial sensory activation. Advertisers do this so that that can satisfy a regulatory requirement to mention terms and conditions. But some of the terms and conditions might undermine the sales pitch, so having them speeded up will still satisfy the requirement but will also reduce the likelihood that the listener will remember what was said. This is good for the advertiser, maybe not so good for the consumer. For another example, something I remember doing when I was a kid and that I also saw my own kids doing, I'd be trying to count something (money, game pieces, trading cards, etc). I'd have to use my inner voice or even count out loud. My younger brother would try to throw off my counting by shouting out other numbers. is that I'd be counting '14, 15, 16, 17....' And he'd be yelling '22, 7, 34,' or '2, 4, 6'. I'd lose my place because the information that I was trying to process with my phonological loop (the counting) would be pushed out by the information in my acoustic store (the numbers I was hearing).

Most of the examples I've been discussing are about verbal processing and verbal memory, but working memory is not just a verbal phenomenon. We experience the world primarily through vision. We see what's in front of us and we can use an 'inner eye' to imagine things and also to hold the visual representation active as we're processing it and other things in the visual field. This is a visuospatial working memory, which Baddeley terms the 'visuospatial sketchpad'. Visuospatial working memory operates in an analogous way to the verbal working memory, except the assumption is that the visuospatial working memory system recruits visual and motor areas of the cortex in order to maintain representations with primarily spatial characteristics. Like verbal working memory, this visuospatial working memory also seems to be highly correlated with thinking ability.

One of the problems that the working memory model needs to solve is that these two systems, verbal and visual working memory, are closely tied to perception and so the information being processed eventually needs to be managed and merged. Sometimes you need to pay more attention to what you see, other times you need to pay more attention to what you hear, and sometimes you need both. In other words, the systems need some kind of control. Psychologists usually call this 'executive function'. And this part of the system functions like an executive function should function.

In Baddeley's standard version of working memory, the central executive (that's Baddeley's term) coordinates resources between the two other subsystems. But other theories of working memory place a greater emphasis on the independent operation of the executive functions. What are these executive functions? Baddeley's theory, and other theories, highlight a set of general cognitive functions such as task switching, inhibitory resources and selective attention. The central executive in working memory is related to attention and seems to reflect our ability to pay attention to the activity in these working memory subsystems.

If you want to excel at intellectual tasks or cognitively demanding tasks, like studying chemistry, learning to code or understanding financial markets, it can really pay off to understand how these executive functions work. They are what really make your working memory work. The executive functions are what most of us think of when we think of doing mental work. They are how we control our thoughts and behaviour, and how we make our memory work. These are important functions. Let's take a closer look.

One of these executive functions is commonly called 'task switching'. Task switching is the act of switching attention from one behaviour to another. Task switching operates at many different levels of cognition and across many domains, which is why it seems to correlate with other measures of cognition and performance. We covered many examples of task switching when we discussed how we multitask and selectively attend. As with any aspect of multitasking, there is always some cost and as you switch from one task to another, you might experience a momentary loss of attention or lose a small bit of what you are paying attention to. That's the cost of task switching. Just like when you switch channels on a television or switch to a different app on your phone, you might lose a small bit of information from the end of the channel or app you switched from and also a small bit of information from the beginning of the one you switched to.

Inhibition is another one of the executive functions and it's related to task switching. Inhibition is a process that lets us ignore something. This is a process that lets us inhibit attention to something, inhibit a response, or inhibit an action. This is a general function that we rely on to *not* do something. This is important. We need to be able to ignore irrelevant perceptual features or irrelevant or unnecessary thoughts or emotions so that we devote more attention to relevant ones. Inhibition is also what we need for higher-order thinking, such as reasoning, making decisions, and testing predictions and hypotheses. For example, suppose you are trying to solve a problem, maybe a puzzle, a game or a mathematical problem. Unless you are lucky enough to solve it on the first try, you will end up trying several possible ways to find the right answer. Each time you test out one possible set of steps to solve the problem, you check to see if it's working. If you find that your current strategy is not working, you will need to go back to the beginning and start over. This requires some inhibition. How? Well, you need to inhibit your attention to the steps that you tried the first time. You need to make sure you *don't* try the same set of steps again. You need to switch to a new solution and inhibit the old one.

Inhibition is critical for these complex cognitive tasks. It's also important for social interaction because you might need to inhibit your first reaction to a stimulus and respond in a more polite way. Inhibition is critical for getting things done. You need to inhibit your desire to check your phone every 2–3 minutes in order to work on something. Inhibition is important for learning to avow some of the mistakes that come from the heuristics and biases that I wrote about in Chapter 6. Although it's natural to rely on the availability of memories to make decisions and predictions, we also know that this can lead to biases and mistakes. In order to sometimes override your first intuition from memory, you may need to inhibit the first thing that comes to mind and allow yourself to consider other options that might take longer to retrieve from memory.

Because executive functions like switching and inhibition seem to play such a large role in higher-order thinking, many researchers have proposed that executive functions are the primary intellectual component of working memory (Kane et al., 2004). The executive functions serve as a general-purpose working memory system and even seem to be the primary determinant of general intelligence. In other words, the lower-level components, such as the phonological loop and the visuospatial sketchpad, may not be contributing to higher-order thought as much as the executive function does. Executive function availability and capacity may be the core determinants of thinking and reasoning ability. From an individual-differences perspective, people with higher executive function abilities may end up be being able to perform better on skills and tests that are associated with intellectual ability, like school and thinking work. In general, higher executive function abilities are associated with achievement.

Conclusion

At the start of this chapter, I mentioned that there were at least two good ways to divide our understanding of memory. Duration (short, medium and long) and content (events, facts, motor action, words and images). The shorter side of memory, this sensory system and the working memory system, reflects the contents of our thoughts and they are closely tied to perception. These systems store information about what we're thinking about and we maintain that information with active rehearsal and a reactivation of perception. But these representations don't mean anything to us unless we can make connections with things we already know. Our working memory system is an intermediary between the world that's out there; the world of sensation and perception and the world that's in here; the world of long-term memories, concepts and knowledge.

CHAPTER 8

Knowledge: The Desire to Know and Explain

Memory is used for more than just remembering where you put your keys, a phone number or a list of words. Short-term memory is the kind of working memory that lets us think about things, consider ideas and understand language. Working memory helps us to create a stable representation of the world. But of course, we can only create a stable representation if we have some idea of what we are looking at. We can only know what we're looking at and listening to if we have some representation in memory or a concept. Working memory may be a system of active thought, but it depends on being able to merge what we know with what we are sensing and perceiving. But how do we actually know anything? How do we create a representation in our mind and in our brain that is a record of something that already happened? How do we *store* a perceptual experience so we can re-perceive it?

Chapter 6 was about how we use memory for thinking in general, how the errors and heuristics come from the same place. Chapter 7 was about short-term and working memory. This chapter is about the rest of it. All of it. This chapter is about how and why you can remember anything and how you can remember just about everything. This chapter is about your long-term memory and knowledge.

In Chapter 6, I hinted that there are at least two ways to think about your long-term memory and knowledge. We have memory for things that happened to us. Any kind of thing, important or mundane, like a wedding, last week's dinner or the first day on a new job. These are episodes, personal events that happened to you in the past. These events constitute one kind of memory called episodic memory. But we also have memories for general knowledge or facts about the world. This is called semantic memory. I'll explain both of these kinds of memory and more. I will cover the basic tenets of episodic memory, such as its reliance on a sense of time and its malleability. And I'll also cover theories of semantic memory, such as spreading activation of memory and different ideas of knowledge organisation. I will also include practical information in this chapter about how to remember things more effectively. One of the keys to a more effective and more efficient memory is learning to map new ideas on to an existing structure. By taking advantage of the way knowledge is organised, you can usually improve your ability to recall information. If you want to think better, it can be important to get to know how your memory works, so that you can get better at remembering the things you want to remember and also learn to recognise how and when errors might arise.

Different Kinds of Long-Term Memory

When I discussed short-term memory, I suggested that we divide it into several different and distinct kinds of memory. The divisions were mostly a function of time or modality. The sensory memory system is very short, almost instantaneous and closely tied to perception. The working memory system can be retained a little bit longer and its contents can be maintained through a rehearsal process. Within this working memory system, we distinguished among visually based memory and verbally based memory. What we didn't talk about was the actual *content* of those memories. That's because we made the assumption that the specific contents did not matter. The working memory buffer seems to store memory about books, bagels and baseball in the same way. And with sensory memory, although it preserves a veridical representation of what's in front of you, computationally, the process seems to be the same. You don't have a separate sensory system for cars and another for cardinals.

With long-term memory, however, the contents of the memory are more important and interesting. We seem to organise things according to general knowledge and specific experiences. And within that, we often organise and access information conceptually. That is, we have many memories for the same thing that might be stored in the same way. Similar ideas seem to be stored near each other in a conceptually organised, psychological space. We often have a rich and complex network of facts and memories for things we're very familiar with. If you're someone who enjoys cooking for example, you probably have a lot of knowledge of cooking techniques, cookware, ingredients and recipes. And you might have an extensive memory for specific things you have cooked: like that time you baked the best loaf of bread. Or that time your sauce did not come together the way you expected. It is pretty reasonable to expect that your knowledge of food, your memory for cooking techniques, and your memories for specific cooking experiences will be much more well developed than memories in some other domain. If you have very little interest and experience in motor racing, for example, then that concept is not going to be as well developed. It is not going to take up as much psychological space. And you might have few (or no) remembered experiences. If you have never been to a grand prix, you would not have anything there to be a part of your memory.

Earlier in Chapter 7 when I was discussing the role of memory in identifying birds at the birdfeeder, I drew the distinction between memories that are organised facts and memories that are organised around events. These fact-based memories are usually referred to as *semantic memories* because it's the semantic and conceptual content that seems to be the most important thing. The event-based memories are usually referred to as *episodic memories*. These are memories for specific episodes. Specific events that happened to you or that will happen to you. These systems seem to operate in different ways. There seems to be a functional difference to the ability to remember and know things in general and the ability to recall and recollect things in particular. The functional needs are different, and these systems seem to help you achieve different goals. But they are not separate. They interact. They overlap. They influence each other. It's easy to see this with an example, in this case an example where specific events can eventually result in a new concept.

Stories shape our memories

Many years ago, when my younger daughter was first learning to speak, we had an experience that illustrates the interaction between semantic and episodic memory. You have probably had many similar experiences in your own life.

For me, this one stands out. There's something about watching a toddler learn to speak, learn to observe and learn to form concepts, that's deeply satisfying.

One day, we took my vehicle in for an oil change at the car dealership. It was a new vehicle and might have been in for one of its first scheduled service visits. In this dealership, you could sit in the waiting area and watch the progress on your vehicle through a window. We were in for a basic oil change and maybe a tyre rotation, so this meant putting the car up on a lift, draining the oil, and moving the tyres around. My daughter was only a year and a half old, and so this was a new experience for her. Pretty much everything was a new experience for her. Like most toddlers, she was fascinated with everything. Big things, small things, things that were interesting to anyone, and things that were probably mundane to anyone but a toddler. So, there were lots of new things going on. It was a new place, a new experience, and she was a one-and-a-half-year-old sponge with a lot of energy and a desire to figure things out and form new memories.

She did the usual stuff, flipped though some of the books there and had a snack but when she walked over to the window where you could see the operations, she stopped and stared with her mouth open. Seeing the car up on a lift was *really* new. I think she was blown away; she was completely blown away with what she saw. I still remember how her eyes got wide when we watched our car being put up on the lift. It was our car. Our huge and substantial car that we drove everywhere in. And it was up in the air. She's only ever seen it from one angle. And that angle was not this angle. This was, as far as I can tell from my own intuition of what she was thinking, a truly mind-blowing experience.

Now that was not a mind-blowing experience for me. It was common. It was a chore. But her reaction made it special to me also, which is probably why I remember it to this day. But I wasn't really paying attention to committing all of it to memory (I'll explain later how and why I seem to have a strong memory for this event).

For the most part, I did what most parents do. I talked about what was happening. I said something like 'the car is going up'. She repeated the word 'up'. And I probably said, 'yes that's right, it's going up'. After some more repeating of the word, and some more reinforcement from me, she settled down. But she was still transfixed by the sight of our car on the lift. Then, the car was lowered. Wow. This really caught her attention. The car went down. Maybe she was expecting it to stay up on the lift forever? Who knows, but for whatever reason she was even more excited to see it come down. I think I said something like 'now the car is down' and of course she repeated 'down'. She repeated that a few more times.

Then the car went up again. I think that was almost too much. I thought she was going to break through the window into the service bay. I think this might have been the most exciting day of her life so far. Not that she had much to compare it to, being only 18 months old. But in the span of a 30 minutes, she saw the car go up. She saw it come down. She saw it go back up. I'm sure at that point she was on the edge of her little tiny seat, just wondering what would happen next. You can just guess . . . it came down.

That was it. For her it seemed to be the best day ever. She yelled 'down' again. Laughing. And the other people were kind of laughing too. She had a great time. But she also learned some new things. She learned the words up and down. She learned that there is a relationship between the word and an event in the world. She learned that an 'up' event might predict a 'down' event. She learned that 'down' might predict an 'up'. Even in that simple and commonplace experience, she learned a new idea. She learned the words and concepts for 'up' and 'down'.

Before I get too carried away with my story, I want to connect these events to some cognitive psychological concepts about memory. And before I get too clever, I also should point out that connecting events to concepts is also the entire point of this story.

My daughter had certainly seen things go up and down before. This was not the first time she heard the terms, because we would have asked her 'do you want to get down' when she was in her cot or chair. She had heard the words. What she was doing was mapping the term to what had just happened with the car and connecting that association with other times she has seen and heard 'up' and 'down'. She was forming a new association (the car and the word) and connecting it with previous examples. This is an example for forming new semantic memories, because the association between the word and the action is semantic by definition. 'Up' means a thing goes up. 'Down' means the thing comes down. Up and down are often connected with other. They are related and predictive. These are all concepts related to the meaning of the words and the events that are bound together by these words.

These are semantic memories because the memory is for the semantic content. These are semantic memories because what matters is the meaning in general and not the specific event. In fact, for most semantic memories, being too specific would be detrimental. It would not be helpful for her knowledge of up and down if she thought it only applied to being at the car garage and only to cars. She needs to learn that up and down are general concepts that apply to all sorts of objects and events. Semantic memory creates and stores those general associations and does not pay too much attention to the specific event as part of the memory. Semantic memory seems to try to create generalisations at the expense of not remembering the specifics. But we do remember specifics, sometimes. How does that work?

The first part of this story on memory is about the encoding and storage of semantic facts. About learning up and down. But there is more to my story. The second part of the story is about remembering the event. At the time, my daughter was attending a part-time day care that was connected to the university. After the car garage, we drove to the day care centre, and I went to campus. A few hours later, I picked her up and we headed home. We drove past the car garage where we had done the oil change earlier in the day and I pointed it out to her. Her eyes lit up and she started going on about 'up' and 'down' again. She was saying up and down, as if she remembered the event from earlier. As far as I can tell, she was reliving some of that episode. She recognised the experience as a familiar one. She recognised the experience as something that has happened. She had formed a new episodic memory. She also remembered the event when we drove past. There was a real event that was connected to the words and she recognised that she was part of that event. Unlike the semantic memory, which is a memory for the general idea of 'up' and 'down', this episodic memory is very much concerned with recording the specific episodes. Episodic memories are memories for personal events. We try to store specific information and we connect the specific information with ourselves being there. In this case, she recognised that she had been there before, and she told me that through her remembering the concepts that she had learned in that episode.

In this story, we see the strengthening of a semantic memory (the general concept of up and down) and the creation of a new episodic memory (the recognition that she had seen the car go up at that place earlier in the day). These are two different kinds of memories because they are storing two kinds of things, even though they may be related to the same original event. These are two kinds of memories because each keeps track of a different aspect of the event. The semantic memory is general. It creates new links with existing concepts. But the episodic memory is specific, and it allows the mind to store different information.

Fuzziness and flexibility

When we drove back to the garage, and she recognised it, did she say 'up', 'car up', 'car go up'? I have no idea. Despite remembering a lot about this story, and despite my having a memory for the event in general, I don't remember many of the specific details. Come to think of it, I don't remember a lot of what should have been important details about that episode. I don't remember exactly how old she was, what day it was, what time of year it was. Why? Those things were not really important at the time, so I did not make any attempt to commit them to memory. If I do not remember those details, how is it that I remember this event at all and how accurate is my own episodic memory?

Well, I told the story in a class shortly after. I was teaching my class on 'Cognitive Psychology' and I used this story to illustrate the difference between semantic and episodic memory. I told the story to the class, much like I am telling it here. It's a good story for this. It took an event, an episode that happened to me (seeing my daughter) and allowed me to talk about her episodic memory and her semantic memory. I can remember it now, because I remembered it then. I remember it now because I elaborated on it and talked about it in class and in subsequent classes. Which means that I strengthened that memory. It also means that in the elaboration and telling, I may have added or elaborated upon a detail or two. But those have now become part of the memory. The memory is not exactly a perfect mirror reflection of the event that happened. Instead, it's a reflection of that event *and* of the retelling. It's all but impossible to know which of the minor details were directly experienced by me when I first saw the event and how many were sort of added by me later.

One of the curious things about episodic memory is the way in which we experience it. This experience is different from the fact-based semantic memory. When I remember a fact, I generalise and recall or recognise the information at a pretty abstract level. You often expect the details to be fuzzy. In fact, fuzzy might even help. It's the fuzziness that helps that 'filling in the background' process that I talked about in Chapter 6. It's the fuzziness that lets you fill in the details of any situation from your memory and knowledge so that you don't have to keep trying to perceive and recognise every single detail in a scene. For example, if I have some general knowledge about oil changes and car maintenance, it's probably good if I just rely on that information to fill in details of the story. It's general knowledge. General knowledge needs to have some flexibility. We don't even notice the fuzziness in part because it's so helpful.

When I recall an episode, however, flexibility is the *opposite* of what I want. I do not want to have some fuzzy and flexible memory for what happened. If fuzziness and flexibility help in semantic memory, they seem to hurt in episodic memory. So why have the flexibility? Well, I don't think we actually have a choice. The problem is that each recollection becomes a new episode. Each retelling and each recounting of the event *is* a new event. So now I have a memory for the original event and a recollection of that original event. If we assume that these episodic memories pass through and are active in the working memory system, then we expect the memory for the original event to activate working memory and that activation can itself then be stored. In other words, the act of remembering creates its own new event that is connected to the original event and to previous remembering. If I remember something slightly wrong or slightly differently, then that becomes part of the memory. If I embellish the story, then that embellishment passes through the working memory systems in the same way as retrieved information does, and so there's a good chance that it will become part of that memory. It's inevitable. The next time I remember that event, I might remember the event with the new detail that I added. And it becomes something that I remember. It becomes something that is reinforced and strengthened.

It is easy to see how your memory can be distorted over time. This distortion comes from the same place as the helpful fuzziness of the more general knowledge. Memories are fuzzy whether we want them to be or not. When this fuzziness helps, as in filling the background details, we never notice it. When this fuzziness hurts, as in adding or deleting details from an episodic memory that you expect and believe to be an accurate reflection of something that happened, we may not notice it either until someone points out that our memory is not accurate.

How can you learn to benefit from the fuzziness sometimes and learn to avoid the errors other times? One solution is to try to avoid elaborating on a story if you want to preserve the core tenets. This solution is not very practical though. Most of us like elaborating when we tell stories. That's what makes stories fun and interesting. That's also the goal of many stories: to entertain, not to help form accurate and veridical memories. It is much easier to say 'try to avoid elaborating' than to actually do it. It's not easy, because fuzziness and elaboration are an inherent part of how memory works. It's impossible to avoid. It is possible to be aware, though, and to use that awareness to decrease the likelihood of a mistake. You can learn to be aware of the nature of your memory. Are you remembering a story or a fact? What are the goals of the memory in question? Are you adding new details? Psychologists call this metamemory. Metamemory is knowledge about one's own memory. If you want to improve your metamemory, it can help to know more about how your memory works at a fundamental level.

With that goal in mind, let's spend some time learning about memory systems, how they work, and why they work the way they do. The next section is less about stories and more about the mind and the brain. I'm going to base this discussion on three kinds of information that we store in our long-term memories and use to guide our behaviour: memory for general facts (e.g. what are shoes), memory for personal past (e.g. can you remember when you last bought a pair of shoes), and memory for motor procedures (e.g. how to tie your shoes). Psychologists refer to these as semantic memory, episodic memory and procedural memory. These are not the only ways to divide and describe memory, but this distinction is one of the most useful and widely supported.

Declarative and Non-Declarative Memory

In our discussion of short-term and working memory, we divided things by modality, that is, sensation and perception. The working memory system is a representational buffer that sits between the outside perceptual world and the inside mental world. Working memory has no choice but to be organised this way because of how closely the system is connected to perception.

It is tempting to imagine that our long-term memory is organised in the same way, according to perception. And indeed, we do have perceptual experiences that are part of our long-term memory. But as we saw in the story about the car going up and down on the lift, many of our long-term memories are primarily organised by content. We seem to have some memories that are mostly factual in nature and others that are mostly organised around personal events that occurred in the past. Sometimes these overlap as we saw in the previous section. But these kinds of memories are not exactly the same. They do not always overlap.

Our brains store and consolidate states of activation into memories that can be retrieved to guide behaviour. Some of these stored states of activation are for facts and events that we are clearly and explicitly aware of. These are memories for things we know that we know. Memory for our name, memory for the names of birds that land at the birdfeeder. Psychologists usually refer to this as 'declarative' memory. Declarative memory comprises your system of 'knowing what' something is, as opposed to 'knowing how' to use it. These are memories that you can declare the existence of, whether that's your mother's name, the name of your street, you favourite food, your first date or the last time you went on holiday. You can inspect your memory and recall or recognise that this information is part of your memory or part of your knowledge. Now of course, you don't always have to be explicitly aware of what's in your declarative memory system. But you would be able to retrieve it if you needed to. And perhaps most importantly, these are memories that we can talk about using our language.

But your memory is much more than storage of things that you can discuss, describe and declare. You have non-declarative memories too. Non-declarative memory comprises your system of 'knowing how' and includes things like how to grip a coffee mug, how to write your name, how to tie your shoes, and rules of grammar in your native language. For most of these things, you still know that you know them. But you can't inspect them. You can't really describe their contents or meaning using language. You can try but even that is not a good representation of how these memories work. Exactly how do you store and activate the information for gripping a coffee mug? You don't think about that when you actually use the memory. You just use it.

These two systems, declarative and nondeclarative memory, are the two broadest distinctions in long-term memory. As with a lot of things in psychology there are other ways to talk about this distinction. These systems are sometimes known as the 'what' and 'how' systems because we seem to have one kind of memory to remember what things are and another kind of memory to remember how to do things. These systems are also known as the explicit and implicit systems because we can remember some things overtly and explicitly but other things without even realising it. These systems are also known as the verbal and non-verbal systems because we store and access some information by way of our own language – an inner voice – and other things by way of non-verbal actions and behaviours.

And within each of these systems, even more divisions may exist. The proliferation and subdivision of memory into memory systems has been a long-standing concern for some psychologists. On the one hand, we clearly have one implementation: a neural system. All these memories are stored in the brain as states of activation and as connections among clusters of neurons. This is a single biological system. But on the other hand, these different kinds of memories operate in different ways at a cognitive level. They seem to operate at a different neural level as well, because some memories will rely on verbal areas of the brain (in the temporal lobe) and others will rely on motor areas in the parietal lobe. Do we have one memory system or many memory systems? And if there are multiple systems, how many? Where we draw the lines has been one of the central topics in memory research for decades. I like to think about memory as declarative and non-declarative because it's intuitive and supported by decades of research.

So, let's look at these memory systems in more detail. I want to begin with declarative memory, because my earlier story about the car and learning the 'up' and 'down' illustrates the two kinds of memory within that declarative system. Then I'll cover the nondeclarative system.

Declarative memory for facts and events

Your declarative memory is explicit, verbal, and includes all the things you can describe and declare. It can be divided up further into memory for general facts and memory for personal events in the past. The distinction between memory for general facts and knowledge, and memory for personal events was explored in detail by a Canadian psychologist from the University of Toronto named Endel Tulving. Tulving is one of the founding pioneers of cognitive psychology and cognitive neuroscience. He began teaching at Toronto in the 1950s and remained there until retirement in 2019. He spent much of his professional career developing and exploring theoretical models of memory.[28]

In his original treatise on memory systems in 1972 (Tulving, 1972), Tulving distinguished among the different 'categories' of memory that had been studied up through the late 1960s: active memory, auditory memory, short-term memory, working memory and long-term memory. These are more or less the same divisions that were described by Ebbinghaus in the early twentieth century and more or less the same divisions that I've been discussing here a hundred years later. He also refers to what was then a newly described system of knowledge organisation called semantic memory. Prior to Tulving, most theories simply drew a distinction between short-term memory (the working memory we discussed) and long-term memory, which was everything else. The assumption was that anything you wanted to remember for more than a few seconds could be rehearsed in short-term memory until it was able to be stored in long-term memory. But in the late 1960s some researchers, often working with computer models of knowledge storage, realised that much of our long-term memory was organised by semantic content. In other words, not all long-term memory is the same. Memory for facts, memory with semantic, *meaning-based* content, was organised in a way that reflected the actual content. This semantic memory system, which is included in Tulving's concept of declarative memory, is a system of memory for facts. These are facts we know, places, names, concepts, colours, cities, animals, plants and food. Semantic memories have meaning. Semantic memories have labels and names. Semantic memories form a densely connected network in your own mind, and in many cases, the meaning is shared with other people by way of language and common knowledge. As we'll see soon, this is a very language-based system, unlike the procedural memory system I discussed above.

Semantic memory

Semantic memory is usually thought to be organised conceptually. My daughter's concept of up-and-down reflects the beginning of her semantic memory for those concepts. She would later be able to connect that example

---

28 In addition to Tulving's theoretical contributions, he has also been a graduate advisor and many of his students have been influential figures as well, including Daniel Schacter, whose work I discussed in Chapter 6, and Henry Roediger, whose work I discuss later in this chapter

of up and down with other examples: toys, foods, the cat, even her. Very general. And that's the important thing about semantic memory. The organisation is key. It's the organisation that affects thinking.

Most theories of semantic memory organisation assume that the organisation of thoughts and ideas in memory reflects the organisation of things in worlds. If two things are similar to each other in the outside world (the world that you are perceiving), then they should be similar to each other in the perceived and remembered world. In this case, similar means they will be stored in a way that makes them seem close. When you think of one thing, such as 'bread', it should be easy and natural to think of something related, such as 'butter'. This is called a semantic distance. Similar things are close to each other in psychological space and therefore they have low semantic distance. Dissimilar things, or things that are not connected, can be far away in psychological space. They therefore have a high semantic distance.

This is more than just a metaphor. This is a metaphor with some predictive power and the idea of semantic space is an important assumption in psychological theories and models. Let's use a simple example. In the supermarket, different kinds of apples are sold in the same area of the store. If I stop by the green Granny Smith apples and then decide to buy red gala apples, it's going to take only a few seconds because they are near each other in the store. The dishwashing detergent, on the other hand, is in a different place. It takes longer to get from the green apples to dish detergent than it does to get from green apples to red apples. Distance predicts the amount of time it would take. Our memories are often retrieved in the same way. And theorists and psychologists have tried to understand how and why this is the case.

The grandparent of semantic memory organisation is a hierarchical theory first described by Collins and Quillian (Collins & Quillian, 1969). This theory was developed in the 1960s, during a time when computer science was just coming of age. Computer memory and storage was severely limited by our modern standards. One of the challenges faced by computer scientists was how to structure information to maximise the amount of information stored while minimising the physical space needed to store it. Ross Quillian developed a hierarchical data structure that was able to store propositions without taking up too much space. The hierarchical structure imposed on knowledge representation results in an efficiency. Knowledge is organised in this system as a hierarchy within a spreading activation system. Spreading activation is an idea inspired by actual neural architecture. When one area of a semantic network is activated (by perception or by thinking about it), the activation spreads to other areas and activates them too. The more closely related two ideas or concepts are, the more quickly the activation spreads.

In Collins and Quillian's model of memory, individual nodes represent concepts and facts. These would correspond to bits of information stored in your semantic memory. These nodes are linked to other nodes in the way that ideas and facts in your memory are linked together. The links between these nodes represent different relationships between concepts. Figure 8.1 shows an example of how concepts and facts are connected to each other in a structured, hierarchical way. One of the things that's most important here, and the key insight of Collins and Quillian's approach, is that within this hierarchy, any attributes of the higher-order node (e.g., ANIMAL) are also true of lower-order nodes (BIRD, FISH, etc.). The subordinate facts and concepts inherit properties of the superordinate nodes.

A Semantic Memory Hierarchy

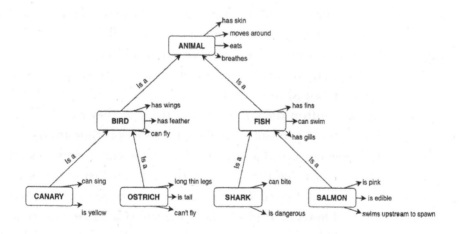

Figure 8.1: An example of a semantic hierarchy. Each node inherits properties from the nodes above it. CANARIES are BIRDS and so they also have the BIRD features (as well as ANIMAL features).

If this system needs to store some facts about canaries (e.g., 'A canary is a yellow bird that sings') it stores only the uniquely canary facts with the CANARY node (yellow, sings) and has a pointer to other levels. With this pointer ('a canary is a bird') it has access to the BIRD features ('feathers') and also other pointers ('bird is an animal'). Other facts ('An ostrich is a bird that cannot fly') can use the pointer to the BIRD node to access some of the same information. That way, facts about birds only have to be stored once, and can be reused and recycled multiple times. The system implicitly stores these complex facts about canaries and ostriches as a distributed hierarchical network. The activation spreads in this network from one node to semantically close concepts faster that it does to semantically distant concepts. These two assumptions, the hierarchical organisation and the spreading activation, allow the model to make some specific and testable predictions about thinking patterns and behaviour. According to this model of the mind, when you think about a canary, 'yellow' should come to mind quickly and easily. But features connected with other levels (like 'has skin') will not come to mind as quickly because it will take longer for the activation to spread to that region of psychological space. It's a model that was designed for computer memory storage, but it also makes predictions about human behaviour.

Collins and Quillian realised that their model was efficient, and they suspected that it could be a plausible way that the human mind is organised and a good model for how we store information as well. They set up an experiment to test this idea. They asked research subjects to complete a sentence-verification task. In a sentence-verification task, subjects are given a statement and are asked to verify if the statement is true or false, and they are asked to confirm the statement as quickly as they can. In these tasks, the important dependent variable is the reaction time to say yes or no. That is, how fast can you respond. The statements themselves are straightforward enough that the subject in the task can answer yes or no correctly. For example, if given 'Canaries are yellow', subjects can answer yes very quickly. However, if given the statement 'Canaries have skin', subjects should still be able to answer yes, but according to the model, it should take longer because activation needs to spread to the BIRD and then ANIMAL level before the property can be verified. The semantic distance between 'canaries' and 'sings' and 'canaries' and 'skin' should be reflected in the difference in reaction time to respond. Their research bore this prediction out. The time to confirm true statements was predicted by the distance between the nodes in a hypothetical semantic network. We access semantically close properties more quickly than semantically distant properties. The structure of our sematic memory representations reflects the organisation of many natural concepts in the world. And as a result, this hierarchical structure affects behaviour. This hierarchical structure affects the way we think about concepts, the way we think about the properties of concepts, and the way we answer questions.

This semantic network is a powerful approach. We see evidence for semantic organisation and hierarchical structure in closely controlled experiments like the ones Collins and Quillian carried out. But there is also evidence for semantic organisation and hierarchical structure in the way we as humans organise things in our built environment. You can find semantic organisation and hierarchical structure in how shops are organised. Similar products are organised into groups and those are organised into superordinate groups. And similar shops might even be located near each other. There is semantic organisation and hierarchical structure in the way the internet is organised. One of the earliest internet search companies, Yahoo, was originally designed to be a hierarchical guide to the early web, and the name was a backronym for 'Yet Another Hierarchically Officious Oracle'. We even organise our libraries this way, by putting conceptually similar books in the same area of the stacks. These things and places are organised semantically, conceptually, and there is often a hierarchical structure. This is a natural way to organise information. We expect it, and we behave as if things are organised this way.

The basic hierarchical model is not perfect, though. A strict hierarchy does not deal well with typicality. When people are asked to verify properties about 'robins as birds', they may respond more quickly than when asked to verify properties about 'penguins as birds'. This is because robins are more typical birds. They have more connections to other birds and share more features with other birds compared to penguins. Penguins seem to be in their own category, even though we know they are also birds and also embedded in this same hierarchy. In order for a hierarchical account to accommodate this typicality effect, it needs to include some additional assumptions for strength of connections between nodes. Stronger connections are faster.

These concepts – spreading activation and semantic networks – underlie most of our knowledge. Understanding how they work can be an asset. Understanding how they work can help you to improve your own memory and also help you to understand why you sometimes make memory errors of the kind outlined in Chapter 6. Improved memory and many common errors are often the result of the same processes: semantic network and spreading activation. Let's look at an example of each.

Around the same time that Tulving was developing his ideas on the nature of declarative memory and not long after Collins and Quillian published their work on semantic memory, Fergus Craik along with Endel Tulving (Craik & Tulving, 1975), began looking at the role of encoding on memory. Although the focus of their work was on the way information was encoded during learning, the results of their research are well explained by the concepts of spreading activation and organisation of memory – though that's not how Craik and Tulving described their work. Craik and Tulving suggest that there are different levels of processing that are employed when perceiving information and trying to store it in memory. According to their theory, incoming stimuli (e.g., things that are being processed and learned) are subjected to different kinds of encoding. Shallow processing refers to processing information at the sensory and surface level – visual features, sounds, and any other features more closely connected to perception. Deep processing refers to processing information in terms of semantics and meaning. In general, they argued that deeper processing brought about better recall of the information in subsequent tests.

In a series of creative studies, their subjects were asked to learn lists of words in one of several encoding conditions of increasing processing depth: a structural condition, a phonemic condition, a category condition, and a sentence condition. In the structural condition, subjects were shown a word and then asked to say 'yes' or 'no' to the question 'Is the word in capital letters?' Half of the words were presented in capital letters (TABLE) and the other half were presented in lower-case letters (chair), so the answer was always yes or no. Notice that this question does not require the learner to spend any effort thinking about the word or thinking about what it means. In fact, you can easily do this task with words you don't even know, with non-words, with your eyes out of focus, etc. You don't even need to be able to read the word to answer this question correctly, only to notice the presence of capital or lower-case letters. In another condition, the phonemic condition, subjects were asked to answer the question if the word to be learned rhymed with another. For example, they would see a word ('crate') and were asked 'Does the word rhyme with WEIGHT?' In this case, subjects needed to read the word and think about how it sounded in order to answer the question correctly. This is deeper processing. For the category question, subjects were asked if the word was a member of a category. For example, for 'Is the word a type of fish?', if you saw the word 'table,' the answer would be 'no'; if you saw the word 'shark,' the answer would be 'yes'. Unlike the structural and phonemic condition, you need to know the word to answer the question. It activates your semantic network. Activation will spread as you consider how to answer the question. This is even deeper processing. Finally, in the sentence condition, subjects were asked to answer a question about whether the word fitted a sentence. For a sentence like 'He met a _____ on the street' the answer is yes for 'friend' and no for 'cloud'. Like the category question, this question requires the learner to think about meaning. It might even require the learner to imagine the word to see if it fits. This is also deeper processing. Unlike the structural condition, the learner would not be able to answer this question if they did not know the word.

When the participants were later tested with a memory test, the words that were paired with a category question or a sentence question were remembered better than words that were paired with a structural or rhyming question. Just as Craik and Tulving had predicted, deeper processing encouraged better memory. It seems that attending to the meaning of the word encouraged deeper processing, a stronger representation, and better performance on a memory test. It activated a semantic network and activation had to spread to other areas and levels in order for people to answer the questions. This is a robust result. It has been replicated and extended in a number of different studies and labs. Even when subjects were not told that this was a memory test, engaging in the deeper processing required by the semantic condition still resulted in fairly good performance on the recall task. Other experiments controlled for the amount of time subjects spent processing the initial word, and even when they were made to process the words for longer durations in the shallow conditions, performance was still better in the deeper, semantic conditions. The central message is the memory for information can be strengthened when the information is given considerable, effortful thought.

Except when it doesn't help. There's always a catch.

What does this mean for your memory? When you really want to remember something well, try to relate it to something you already know. Try to elaborate. Try to connect the new thing that you are learning to a strong, existing semantic network. The elaboration will result in the new information having a lot of connections in common with what you already know, and this will make the information easier to learn and easier to remember. Memories are enhanced when they are connected to other things. Elaboration helps.

The catch comes in the form of a very clever paradigm designed to create false memories. Not forgetting, but false memories of the kind where you think you experienced something but did not. Seeking to demonstrate the ease with which false memories can be created, psychologist Henry Roediger explored an interesting paradigm, which is now known as the Deese, Roediger, McDermott paradigm, or the DRM task (Roediger & McDermott, 1995). In the original version of the task, subjects are presented with several lists of words that they are instructed to remember. For example, subjects might see the words: *bed, rest, awake, tired, dream, wake, snooze, blanket, doze, slumber, snore, nap, peace, yawn, drowsy*. Notice anything? The word *'sleep'* is not in this list, and yet all of these words are connected to the concept sleep. In the DRM task, subjects were given several lists of words like this, and were asked after each list to recall as many words as possible. After six lists, they were given a recognition task which included words from the study lists, new words, and the target words that were not on the list but were highly associated (*sleep*). What they found was that people would falsely recognise the target words like *sleep*. Subjects were convinced that the target word had been presented. In versions of the task when subjects were asked to indicate a distinction between whether they remembered the word or whether they just knew the word, most subjects indicated that they explicitly remembered seeing that word.

When I teach about memory in my university class on Cognitive Psychology, we do a demonstration of the DRM task in class each year. Sometimes I do it by showing the words on a screen and I ask students to try to remember the words. Then I show the test words, one at a time, up on the screen and ask people to raise their had if they remember seeing the word. Lots of hands go up when we get to sleep. Now that nearly all students have a laptop or a device, I create a Google form that presents the words one at a time. After a short pause, students see a list with all the test words (old words, new words and new target words) and they click 'yes' if they explicitly remember seeing each of the words. Again, there are a lot of people who falsely recognise the target words. Sometimes students do not believe me that the word *sleep* was not on the list and so we go back to look at the list to prove that it was not there. It is not unusual for students to be surprised that the word is not actually part of the list. It's a very strong sensation.

This is clearly a memory error. The word, though highly associated, was not present. People who indicate remembering it are demonstrating a source misattribution error. Spreading activation among all of the related words results in high activation of the target words, and this is strong enough to result in a false memory. In other words, people think they remember seeing the word *sleep* in part because they really did experience the word. They did not experience it out there in the world by reading it. But they did experience it in the mind as a result of spreading activation. They saw it their mind's eye, and of course they saw all the other words that way too. *Bed, rest,* and *awake* are perceived then activated in long-term memory. They connect with other sleep words. You might even consciously notice that they are 'sleep words'. You might even say to yourself, using an inner voice. 'Hmm, these are all sleep words'. It's all but impossible not to have strongly activated the word *sleep* and to have the subjective experience of having seen the word.

Spreading activation and a semantic network are design features in how your brain and mind organises information. They are unavoidable and usually beneficial. But the same features that can be used to improve memory (as in the levels of processing work described by Craik and Tulving) can also result in false memories as described by Roediger. Semantic memory is great for storing what we know, for elaborating on ideas, and for making connections with concepts. It's not so good for retaining an exact copy of the past. Semantic memory is not designed for that.

But we do sometimes re-experience the past. We sometimes even enjoy reliving the past as a form of entertainment. We relive events. We remember people, and places, and things. Whilst working on this book, I remembered many different things and events that I ended up writing about: a near car accident, a friend's actual car accident, an experience in the classroom, a recollection of a learning moment with one of my girls. Sometimes the experience is like watching a movie. I can see the event play out like it happened (or more accurately, like I *think* it happened). But other times, this recollection is just a simple act of verification. Did I buy coffee at the store or did I forget? What did I have for breakfast? What was the name of the person who took my order at the restaurant? Are these past, personal events any different from the general knowledge systems that comprise semantic memory? Or are they similar in structure function, with the only difference being the content: personal events vs general facts?

Episodic memory

Let's return to 1972 and to the foundational work of Endel Tulving. Tulving's original description of declarative memory also described another, more personal form of memory he referred to as episodic memory. Episodic memory is memory for personal events in the past or even in the future. Episodic memory allows you to recall something that happened to you in the past and to set a mental alarm for something that you expect to happen in the future. It is, as Tulving suggests, a form of mental time travel. Although episodic memory still relies on the same basic neural mechanisms that semantic memory does, it is distinct in its contents and its use. Unlike general knowledge, in which we store and retrieve information about the world as we experience it, episodic memory is deliberate and often used to store our conscious experience of where we are, what we are doing, what we have done and what we will do.

Examples of episodic memory are all around. And once you start looking for them, you can see how they differ from general knowledge (though they still overlap with semantic memory because everything does). What was the first thing you did this morning? Can you answer that question without pausing to engage in mental time travel? Maybe. But, more than likely, it takes a few minutes for you to picture yourself doing something this morning. Not every example requires you to imagine the past. You also rely on episodic memory to keep track of what's happening. What was the colour of the car that cut you up a few minutes ago? What was discussed at the meeting last week? And you also rely on episodic memory to plan and imagine things that have not happened yet. When is your next haircut? Where is your next class being held? Remember that you need to leave the house at 7:30 in the morning to drive to Toronto. What will it be like next year, after we've all adjusted to the COVID-19 pandemic?[29]

Tulving suggests that episodic memory, this ability to engage in mental time travel, is uniquely human. It depends on our inner voice to describe things and a developed sense of self to put yourself in the never ending stream of information. And Tulving suggests that this memory relies on, but is distinct from, semantic memory. He writes, in 2002 (Tulving, 2002):

> Episodic memory is a recently evolved, late-developing, and early-deteriorating past-oriented memory system, more vulnerable than other memory systems to neuronal dysfunction, and probably unique to humans. It makes possible mental time travel through subjective time, from the present to the past, thus allowing one to re-experience, through autonoetic awareness, one's own previous experiences. Its operations require, but go beyond, the semantic memory system. – Tulving 2002, p. 5

I began the chapter with a story about my daughter learning about the concepts 'up' and 'down' and suggested that for her, that was the important information to learn about the event. She did seem to have something like an episodic memory for the event when we drove by the garage, though. She recognised it. But I made a point to remember it. I thought about it, elaborated on it, and engaged my episodic memory to plan to talk about it in class. However, as I also pointed out in my story, episodic memory is susceptible to the effects of elaboration. Episodic memories are connected to the same semantic network. Spreading activation works here too. And because of this, we still make errors.

---

29 I would argue that one of the most difficult things about dealing with COVID and the novel coronavirus is that we are not able to use our episodic memory to plan ahead. When the pandemic first hit Canada in March 2020, I planned ahead to work from home for two weeks. That turned into two months, which turned into four months, which turned into *more* months. In most years, I look forward to the fall semester and being on campus. I can't really do that now. My usual mechanism of travelling forward in time to imagine the fall term is not working because I also know that it's not going to resemble the past. This idea, that the future will resemble the past, is going to come up again in this book when I write about induction.

Remember the DRM paradigm from a few pages earlier? I suggested that the spreading activation caused the word *sleep* to be active and falsely recalled because it was receiving activation from other words and concepts that were presented. The activation is a result of your semantic memory. But the question in the actual task is one that is about your episodic memory: '*Did you remember seeing the word sleep on the list?*' When people make that mistake, it's because they are relying on semantic activation to answer a question about a specific episode that is probably weakly encoded and is also hard to distinguish from the other episodes (i.e. all the other words the person saw).

So episodic memory is not perfect. Far from it. But it is useful for us and it's impossible to imagine being without it. Episodic memory is part of being human. Tulving suggests that it's uniquely human, but that may not entirely be the case. Some aspects of episodic memory are seen in other animals. For example, scrub jays, which are a kind of bird, are known to remember how long ago they cached their food. They would avoid looking for cached food (worms, in this case) when too much time had passed, and the worms were no longer fresh. This may not be the same as recollecting a birthday party or remembering to set an alarm, but it is a form of time-dependent memory.

Non-declarative and procedural memories

But we have a lot more in our memories than just facts, events and all those other bits. Tulving also suggested we have a 'non-declarative' memory system that includes things we can't easily explain or declare. For example, you know how to type: you know where the keys are, how to press them, and how to make them work to create words. When you are typing, the words might reflect what is in your declarative memory system, but the movements that you selected to type are part of this non-declarative system. Psychologists also refer to this as procedural memory. These memories are still a part of your long-term memory, but they are not easy to declare explicitly. They are not easy to describe. Their retrieval and use are often automatic and outside of consciousness. And these memories are plentiful and important for survival. We are constantly using and updating these memories without realising it.

For example, the basic act of driving a car involves non-declarative memory in several ways. For one, you have a memory for the behaviours and actions needed to steer, navigate, accelerate and brake. If you do not drive or have never driven, the same principles would apply to riding a bicycle (or skiing, skateboarding, using a wheelchair or swimming). This use of memory is sometimes referred to as motor memory because the memory is for the motoric actions needed to steer and for sensory motor coordination. These motor memories took time to acquire, and probably involved a lot of practice, but once you learned them, they became easy to retrieve and use. They are now almost automatic. For all intents and purposes, they require little if any consciousness.

Nondeclarative memories are not always strictly motor memory, either, though most have some relation to movement and motion. Let's continue with driving. In addition to knowing how to operate the vehicle, you also need to learn the rules of driving: the speed limit, what traffic signals mean, who has the right of way and what to do when an emergency vehicle is behind you. You probably acquired these through a combination of study and practice, and like the motor actions needed to operate the car, you use them without really being conscious and you would struggle to describe them out loud to yourself or to others.

Driving a car is a complex and dangerous operation. It's so complex and dangerous that we require strict licensing and testing before anyone can drive. But as I've described, many of the memories and abilities needed to operate the car are nearly inaccessible to conscious and verbal inspection. This might strike you as a problem, that you do not have easy and explicit access to the memories needed for a complex and dangerous activity like driving a car. But it's not really a problem at all. It's actually a benefit. Without the need to filter these procedural memories through consciousness, without the need for attention, and without the need for explicit recall, these procedural memories are fast, efficient and automatic. That's a benefit. The cost is that it's not easy to describe to someone else what you are doing. If you do drive, try to remember what it was like to first learn. Or try to remember what it was like to learn a similar and complex behaviour. Or try to remember what it was like to teach someone how to drive (trying to remember these things, by the way, involves the declarative, episodic memory system and shows again how interconnected things are).

Driving a car seems effortless when you are experienced. I learned how to drive in 1985–1986 on a few different kinds of cars and trucks, but I learned early on how to drive a vehicle with what is sometimes called a 'standard transmission'. It's not really very standard now, but this just means that you have to shift gears manually. The driver selects the gearing, and gears are what transfer the power from the engine to the wheels. Low gears maximise power at lower speeds; higher gears maximise speed and you need to be travelling at a sufficient speed to shift into higher gears. A bicycle works the same way. When you shift gears, you need to step on a pedal called the clutch. Depressing the clutch releases the gears so that there is no power being transferred and this lets you shift safely. When you release the clutch, the gears engage. If the engine is running too fast when you try to engage the gears, it can lurch or grind. If the engine is running too slow, you will sputter and stall.

When I first learned to drive, I had the same trouble that everyone else has with the clutch on a manual gearbox. I lurched to a start. I stalled. I would grind the gears. But as I learned to do it properly, those procedural memories strengthened. More practice and more strengthening of the memories, until eventually I could drive a car with a manual transmission. I have not driven a manual car since 2003, an old Ford F-150 pickup that I drove around Illinois as a postdoc. But I know that if I had to drive a manual transmission vehicle again, even eighteen years later, I would have almost no trouble. Those procedural memories are still there. I can still almost *feel* the movements needed to press the clutch, let up on the accelerator, shift, and then let up on the clutch while I press the accelerator again. And although I can relate these steps in broad terms, I can't explain the motions very well. These memories are not verbal. I know they are part of my memory, but I just can't describe them very well.

The nondeclarative memory system is well suited to these kinds of memories because language is not needed and does not really get in the way. This memory system involves areas of the brain that are tied directly to the actions being carried out. In many cases, language is not part of the initial encoding of the memory. And trying to use language to recall or describe things can even impair recall. Try to carry out a simple, well-learned action while you try to talk yourself through it. It's not easy to do.

Final Thoughts

Human memory is remarkable. The computational power of millions and millions of dense connections between neurons in the brain enables us to recognise our friends, choose words for a sentence, drive a car, reminisce about high school, and plan for the future. The structured organisation and spreading activation allow us to elaborate and make connections but can also produce false memories. Being aware of how your memory works can help to improve your memory. Know when to elaborate to strengthen a memory. And know when to be aware of elaboration that might produce mistakes and errors.

Your memory and knowledge underlie everything we think about. Thinking is using memory, using the past, to affect the present and plan for the future. The next two chapters are about how we structure information for thinking and how we use language to manipulate these structured representations. The next two chapters will essentially complete the 'flow of information' metaphor I discussed in Chapter 3. With that metaphor, I described how information flows from the outside world to the brain and mind by way of sensory receptors, how the information mingles with memory and knowledge, and how information activates and updates concepts. In order to think effectively, we sometimes condense and compact information into structured representations known as concepts. We communicate with others (and with ourselves) with language, which condenses information into neat and sort-of tidy packages: concepts and words.

CHAPTER 9

Concepts and Categories

We organise everything we experience into categories and concepts. Everything can be categorised. We represent these categories with concepts. We would not be able to function without concepts. Without a concept, every experience would be unique. Without a concept, we would not be able to recognise things. Concepts are the way we organise the record of experience. Without them we'd be lost. Consider this example: you walk into a large supermarket. Even before you enter, you expect to find a predictable level of organisation. You expect to see a section for children's clothes, a section for hardware, a section for toys, and a section for food and groceries. Within each of these sections, there is often another predictable subordinate level of organisation. For example, within the food and grocery section, products are organised by bakery items, fresh produce, meat, dry goods. And sometimes, things are even sub-organised further by brand, function and ingredient within those divisions. You expect to find frozen pizzas together (a taxonomic grouping) and you may also find pasta sauce and pasta near each other (a functional grouping). I described a similar example in Chapter 8 – how the physical distance between the green apple and the red apple was shorter that the distance between the green apples and the detergent. I suggested that we organise our mental experience in the same way, with semantic distance as a proxy for physical distance. With a stable concept, we can predict where things will be. We can expect all stores to be laid out the same way. That expectation reflects the conceptual structure.

These divisions are predictable because most stores use similar grouping. These divisions help shoppers to know what to expect and where to find things. If you were looking for frozen carrots, you would expect to find them near frozen peas. If you were looking for pickled artichokes, which might be an ingredient you never purchased before, you might try the section where pickled foods are. And if they are not there, you can make inferences about other places they might be and would not be. Stores are organised conceptually. The conceptual groupings are designed around the kinds of products being sold and how they are used. The groupings make sense to us and make it easier for us to find what we need.

Your mind works the same way. Your thoughts and memories, although represented at the neural level by distributed connections of neurons, are organised conceptually. These concepts reflect the organisation in the world (dogs and cats are natural groupings) as well as functional relationships (bread and butter go together). We've covered some of this already, in Chapter 8 in the section on semantic memory. But the study of concepts and categories is focused on how memory is organised by and for thinking and action.

We store and retrieve memories; but we think with concepts.

What are Concepts and Categories?

Concepts are a way to organise our memories and to organise our thinking. They provide structure to the mental world. We rely on concepts to make predictions, to infer features and attributes, and to understand the world of objects, things and events. The study of concepts, categories and thinking is one that emphasises how categories are created and learned, how concepts are represented in the mind, and how these concepts are used to make decisions, to solve problems and to drive the reasoning process. Concepts are at the centre of human mental life because they allow us to consolidate many experiences into a single representation.

I think a diagram could help here, to show how concepts consolidate experience. Figure 9.1 (page 204) shows a hypothetical arrangement among lower-level perceptual responses (perception, attention, working memory), structured representations (concepts in semantic memory and connections to episodic memory), and higher-order thought processes (actions, behaviours, decisions and plans). This is an abstraction of the more detailed 'flow of information' I have been discussing since Chapter 3. At the lower levels, information has not been processed and is essentially in a raw, primitive form. Primitive representations are the features that we perceive. These features – edges, colours, phonemes – receive input from the sensory system (the retina, the cochlea, etc.) and are stored in the sensory memory systems that I discussed in Chapter 7. But in order for us to be able to plan and make decisions, these primitive representations need to be processed and structured in some way. In the previous two chapters, I discussed working memory and declarative, long-term memory. But concepts provide more abstraction. Concepts are mental representations that have a significant degree of structure. Concepts include thoughts and ideas that are similar to each other that share activations and overlap at the neural level. Concepts have sufficient structure and coherence to affect prediction, inference and utility.

Concepts and Categories

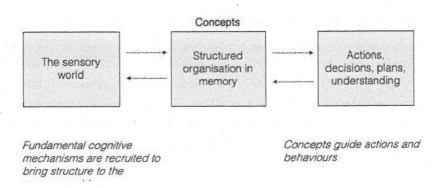

Figure 9.1 A hypothetical arrangement that shows the role of concepts on other thinking behaviours. The outside, sensory world is structured according to features, similarity and rules. We use this conceptual information to make decisions, to reason and to solve problems.

I'm using the terms concepts and categories interchangeably. These two terms appear together often and sometimes seem like synonyms. They are similar, but they are not the same thing. They are not synonymic. I use the word *category* to refer to the objects, things or events in the outside world that are structured into groups. I use the word *concept* to refer to the mental representation that refers to a category. Categories are groupings of things, natural or otherwise, that exist outside the mind. Categories are things that belong together. Concepts on the other hand, are representations. Concepts are abstractions. Concepts exist in the mind and are how we represent categorical groupings. Sometimes the concept reflects the category fairly well, but this is not always the case.

If you are near a computer or have your smartphone nearby, do a Google image search on the words 'coffee mug' or 'coffee cup'. What will come up is image after image of coffee mugs of different colours, design and shape. There will be the standard mug and the taller, travel mug. But they are all readily and instantly identifiable as coffee mug. Some might be more obvious than others. Some have more standard features than others, and others have novel features, such as slogans, logos or unusual handles. But you can still recognise that these images form a coherent category (and Google image search is making the same assumption).

When we learn to classify these objects and to recognise them as members of the coffee mug category, we learn to ignore some of the idiosyncratic and unique features and rely only on the most typical and predictive features. It sounds pretty simple, but this process is not as straightforward as it seems. Many of the most common features, such as having a cylinder shape, will also be found in members of other categories (e.g., a drinking glass, canister, jar, etc.). We might also notice that even those features which are strongly associated, like a handle, are not necessary for being a member of this category. Travel mugs don't usually have handles. And even some standard mugs do not have handles either. Coffee mugs should be a simple, straightforward category, but there is still a fair amount of complexity and variability. Despite this, most of us have formed a reliable concept of what coffee cups are and we are likely to have very little difficulty making a classification decision quickly and easily.

With something simple and straightforward like coffee mugs, the occasional ambiguity does not seem to matter too much. As long as you can put your coffee in it, it should qualify as a member of the category.[30]

---

30 Even the function of holding coffee is not a perfect predictor of the coffee mug category. Many people use a coffee mug for pens and pencils on the desk. This is a coffee mug that does not hold coffee. And many people, my two daughters for example, routinely drink their iced coffee out of a mason jar. This is not a member of the coffee mug category, but it's doing the job of holding the coffee.

But ambiguity in categories can have real and serious consequences. If you choose the wrong category or the wrong concept, you run the risk of selecting the wrong behaviour or making the wrong decision.

As an example, consider paracetamol, the pain reliever commonly found in Anadin and other over-the-counter headache and cold medicines. If you are still at your computer or have your smartphone nearby, do another Google image search on paracetamol. What do you see? It should be image after image of paracetamol tablets, Anadin or similar drugs. If you were asked to say what category you thought paracetamol was a member of, you might say 'medicine' or something like that. And from that you might activate a general concept of medication and infer that it's helpful and beneficial. It reduces headaches and provides pain relief. A lot of related concepts probably come to mind.

You would probably not say 'poison' and would probably not activate a concept of toxic things. But paracetamol toxicity is very serious. Every year many people die from taking too much paracetamol. In fact, paracetamol (or acetaminophen) poisoning has often been one of the most common causes of acute liver failure in the US. It accounts for several hundred deaths a year and causes many more thousands of calls to the Ambulance service. Paracetamol has long been marketed as a safe drug. And if we assume it is safe, we might assume that we can take it without worrying too much. After all, it's supposed to be in category of things that are beneficial. But it turns out that paracetamol has a very narrow therapeutic window. It's safe and beneficial at the recommended dose, but if you exceed that dose, even by a modest amount, it can result in toxicity, hospitalisation, and even death. People often make the incorrect assumption that it is fine to take a little more than is needed. Or if the fever does not go down, they might give their child a second or third dose to help more. And this is exacerbated by the fact that paracetamol is found in many over-the-counter medications (like cold and flu medications), so it may not be easy to tell if you have exceeded this dosage.

Paracetamol is correctly classified as a medication, but that classification could encourage the incorrect assumption that it may be safer than it is. Categorising something allows you to make predictions and assumptions. But in this case, categorising paracetamol as a safe, benign, over-the-counter medication may lead to serious errors. I don't mean to say that paracetamol is unsafe in general. And I don't wish to suggest that you should not take Anadin. Taken as directed, it is a very safe and effective medication. That's why it's so common. But the consequence of categorising it as *always safe* would be incorrect.

Concepts are abstractions of mental and perceptual experience, so there are costs and benefits to representing information this way. A concept is an efficient way to represent most of what is important about a set of things in a category. But because it's an abstraction, it also means that some of the nuance and individuality of the category members is lost. Abstractions like concepts allow people to make quick and usually accurate judgements (e.g. coffee cup or not), but occasionally they come with a cost of misclassification (e.g. safe medication or poison).

Why Do We Categorise and Classify?

People categorise things in part because all animals and organisms have a natural tendency to generalise from prior experience. This idea, known as stimulus generalisation, is present in all species. Stimulus generalisation allows an organism to extend a learned behavioural response to a whole class of similar things. You do this when you treat many different coffee mugs in the same way. I do this if I buy a sandwich from a food truck and expect it to taste like that one I got last week. Your cat or your dog does this when they react to the sound of a can of food being opened up. This kind of behaviour is seen even in the most biologically primitive organisms. In the nineteenth century, William James, one of psychology's founding theorists, noted that:

> . . . *creatures extremely low in the intellectual scale may have conception. All that is required is that they should recognise the same experience again. A polyp would be a conceptual thinker if a feeling of 'Hollo! thingumbob again!' ever flitted through its mind. (James, 1890)*

Setting aside some of James' fanciful terms, it's a clear description of what it means to behave conceptually. When James says 'polyp' he is referring to the tiny, tentacled water organisms that make up corals. Any example will do (worm, ant, snail). And when James says 'conception', he means the ability to form a concept, and not the ability to conceive another polyp. And of course, the polyp in this case does not have a mind that thinks thoughts, but polyps do behave the same way toward some things. They generalise from prior association to produce a response. This suggests that generalisation, and with it the tendency to group memories and experience into clusters, is an inherent aspect of a functional cognitive architecture. These generalisations are guided by similarity. The response rate of a behaviour to a new stimulus will be a function of how similar a new stimulus is to a previously seen stimulus. Through repeated association, the polyp (or worm, ant, or snail) eventually learns to behave the same way towards the same kinds of things.

People and other organisms categorise things to gain a cognitive efficiency to guide behaviour. Forming a concept for a group of things means a reduction in the amount of information to be retained about all the members of the group. A concept condenses many experiences into one abstract representation. This abstraction can be thought of as a behavioural equivalence class. This means that although a group or class of things may be different and many in number, (different examples of coffee mugs), we behave towards all of them in the same way (drink coffee out of them).

For example, my cat Peppermint is like many cats: an opportunist. She is adorably lazy but she's also hyper-aware about her food. She knows the sound of a can of food being opened. If she's sleeping upstairs on the bed or on my desk chair, which she usually is, and someone opens up a can of food in the kitchen, I hear her jump off the bed, patter down the stairs, and trot into the kitchen. Although there are many different cans, and probably some of these cans sound different when being opened, Peppermint behaves in the same way towards each of the sounds. The individual and unique characteristics of the sounds of each can do not matter. Peppermint has efficiently represented a whole *universe* of food-can sounds with a single behavioural response. We do the same thing with our own concepts. We are capable of representing many similar things with one core representation. This is cognitive efficiency. Theories of conceptual representation, which I will discuss a little later on in the chapter, differ in terms of how much information they assume we store in the conceptual representation, and how much information is lost. Most theories of conceptual representation assume that the concept stores general information, with greater cognitive efficiency than many unique individual representations.

The cognitive efficiency that is gained by forming concepts is influenced by the natural structure of the world. People categorise things the way they do because the world of objects and things may be somewhat self-categorising. There are regularities in the world, both physical and functional, and our job as inhabitants of the world is to learn about these regularities. Concepts track and represent these regularities. The idea that there is a natural structure to things goes back at least as far as the ancient Greek philosopher Plato.[31] Plato suggested that we 'carve nature at its joints' when we represent the natural world. Plato was talking about how a hunter or a butcher might prepare an animal for eating. It is easier to cut the animal where the joints are, rather than just hacking it apart. That is why we have chicken breast, wing and drumstick, rather than wings that are cut in half. There are natural ways to cut up an animal. And if you don't eat meat or never thought about cutting an animal into edible parts, the analogy still holds. There are natural ways to peel and segment an orange. There are natural and obvious ways to eat peas or to shell walnuts. Those natural ways exist whether or not humans decide on that. The joints are there. We can theorise that humans form the categories they do based around already existing natural boundaries. We have concepts for fruit, for chickens, and for coffee mugs, because these things are similar to each other in the world. In this case, similar means composed of similar material, similar shapes and being of similar sizes. These attributes and features overlap and cluster whether or not humans decide to recognise that similarity or not. They also segment and discretise whether or not we decide to recognise the discrete groups. As we engage with the natural world, we have no choice but to categorise and conceptualise it along those lines.

If we consider all of these things – the tendency to form generalisations, the possibility of an efficient central representation, and natural groupings that are present in the world – it seems that categorisation is practically inevitable. If categories are an inevitable part of our cognitive architecture, how do they affect our behaviour and thinking?

---

31 This idea probably goes back much further and is also probably not unique to Plato. This philosophical tradition is one of many philosophical traditions that try to explain the world and why we behave the way we do. I'm using it here because it's one I'm familiar with and more importantly, it's one that ties into the precursors of modern psychology.

The Functions of Concepts

Fundamentally, a concept is a cognitive representation that influences how a person reacts, which is why they can be described as behavioural equivalence classes. Just like my cat Peppermint, who formed the behavioural equivalence class for canned food (see the preceding section), the concept can encapsulate experience and drive behaviour. Once an object is classified as a member of a category, we can behave towards it as a member of that category. We see this in many new product categories. Smartphones, for example are not really phones like the phones that many of us grew up with. They are handheld computers. There is no wire, no dial tone, no 'operator'. When you wanted to make a voice call (the only kind) in the 1990s, you would pick up the receiver (the part you speak in) and you'd hear a dial tone. That let you know the phone was connected. As phones changed to cordless phones, cell phones and smartphones, they kept the word 'phone' rather than being called a handheld computer. This creates a continuity with what we know so that we can classify the object, associate it with a concept that we know, and we can predict how to interact with it. And even the newer concepts, like smartphone, activate their own concepts. Pick up a new iPhone or a new Samsung phone and you already know how to operate most of the features because you have an existing concept. Pick up an old flip-phone and you activate a different concept and behave differently towards it. Once you know what category something is in you can rely on your concept knowledge to help you infer things. There are many other examples: a concert that is classified as classical music encourages a different attitude or style of dress than one that is classified as folk music. A wine that is classified as a dessert wine encourages it to be consumed in a specific way and probably not with a steak. Shoes that are classified as trail shoes activate different concepts than shoes classified as gym shoes.

This idea – that categories help you know how to behave or react – might also have troubling consequences on the thinking process as well. This tendency is at the root of many negative racial, ethnic and occupational stereotypes. Because our minds have a tendency to generalise from whatever concepts we have and have formed, we may make helpful as well as unhelpful generalisations. We might (even unconsciously) adjust our behaviour when speaking to a person from our own ethnic group, as opposed to a person from a different ethnic group. We adopt different manners when visiting the doctor than when talking with the receptionist at the same office. We harbour different attitudes towards some racial groups relative to others. Much of the research and literature on stereotyping and racial prejudice falls within the context of classical topics in social psychology and suggests that these categories can bias attitudes and perceptions in subtle and implicit ways.

For decades, the US in particular has had an ongoing problem with violence between police groups, Black communities and, sometimes, protestors. This conflict has spiked at times – in the 1960s Civil Rights era, during the 'war on drugs' in the 1970s, after LAPD officers were seen beating Rodney King and, more recently, after the riots in Ferguson, Missouri and the killing of George Floyd in 2020. When these conflicts occur, there is news, analysis, photos and video. There is also disagreement about the role of police, the role of suspects, victims and protesters. These are all concepts that will have some meaning and coherence for people, but the boundaries and features of those concepts will shift between different people. The concept of 'police officer' means different things to different people. And a person's concept is going to depend on a person's experience. A concept, after all, is a summary representation of individual memories, perceptions and experiences. What would these experiences look like?

One way to imagine is to turn again to Google image search as we've been doing several times in this chapter. You can run a search on 'police officer' and this should return a good assortment of images, many depicting police in action and others in portrait. What is your first impression? Does it fit with your own concept of a police officer? Try a few variations. For whatever reason, if I search for 'New Zealand police' they are mostly friendly seeming images. If I search for 'US police' the images are less friendly. And the images depict a much more heavily armed group. And if I search for 'riot police' there are no friendly faces to be seen. There are not many faces at all, they are obscured by helmets and visors. The world of police officers is still being carved at the joints, but there are different joints and different focus points.

Imagine forming your concept of police on the basis of a few interactions and mostly images of the friendlier, less heavily armed people whose faces are similar to those of your family and friends. Your concept, an abstraction, will include the things that are important to the category in general (makes arrests, carries handcuffs) and maybe some that are specific to your experience (helpful, protects the neighbourhood, looks like people I know). Now imagine that your experience is shaped primarily by seeing police with military style vehicles, heavier weapons, helmets, face shields, and they appear during times of violence, like shootings, riots and conflicts. It's not hard to see how you might form a very different concept in this case. This concept might include the same general features (makes arrests, carries handcuffs) but a different list of specific things based on experience (intimidating, aggressive, uses force). Despite having the same label of 'police', these are not the same concept. People with the former, friendlier concepts will predict and infer features that are consistent with that friendlier concept. People with the latter, more aggressive concept will infer features that are consistent with the more aggressive concept. Both inferences would be a result of the concepts in the individual's mind; concepts that are designed to abstract and compress many experiences into a structured and usable mental structure.

We use concepts and categories to make predictions all the time. These predictions allow us to infer features and also guide behaviour. When an object or an item is classified as belonging to a certain category, we can use what we know about the category to make predictions about other attributes that might not be immediately present, but that we know to be associated with the category. Remember the example of different kinds and shapes of smartphones from a few pages ago? There are different brands, different manufacturers and different models. If you pick up a new, unfamiliar phone (or one of the older models), it may look slightly different from what you are used to, but as long as you know it's a smartphone or a cell phone, you can make any number of reasonable predictions. You know there's a way to change the volume, to make a voice call or to take photos. You probably make some assumptions about how it operates based on your general, category-level knowledge. If you have enough information about its category (smartphone) you can make predictions with some degree of confidence. And because we give names and labels to so many of our concepts, we can communicate the central experience to another. A category label is part of the concept, and it's an effective way to transfer abstract information to someone else. Rather than relate the process by which you discovered how to turn on the new phone, make a call, and take a picture, you can just say 'This is an older iPhone'. Granting the other person some knowledge of the device's category membership gives that other person access to all the category level information. Very effective. Very efficient.

Concepts and categories also play a role in problem-solving. People often engage in problem-solving strategies and heuristics that involve finding the correct solution from memory. Rather than solve a problem by working through the solution, an experienced problem-solver may search for the correct solution concept in memory. Concept-based problem-solving is seen in many domains. For example, expert chess players access store representations for categories of moves. Expert physicians are known to rely on the similarity of patients to previously seen patients, and there seems to be high agreement among expert physicians about how they form concepts upon seeing a patient. In fact, a few years ago, my colleagues (two physicians) and I asked expert physicians directly about the kinds of concepts they activated and considered when they met a patient in their office for the first time (Goldszmidt, Minda & Bordage, 2013). Doctors told us that they framed the encounter by activating concepts on the patient's immediate need, agenda or referral. They activated concepts on prior diseases, concepts about how to manage the patient. They also indicated that these concepts were built up from all their prior experience. When doctors first meet a patient, the encounter activates memories and concepts of previously seen, similar patients and those representations guide the interaction.

I have been discussing these various functions – prediction, inference, communication and problem-solving – and how they are guided and driven by our concepts. Concepts summarise our experiences and help us to behave, act and think like humans. And although these complex mental functions are very much a part of human thinking, they are not unique to humans. James, discussed earlier, suggested that even the most 'primitive' organisms have concepts. Concepts, or at the very least, behavioural equivalence classes, are important for humans and non-humans alike. But concepts and categorisation are also important for machines. For example, most of us are aware that internet companies are working tirelessly to collect data on users, generate trends and to classify things in order to make predictions based on what they think people want. That's why I can suggest a Google image search to you and can know what you'll probably see. Amazon recommends new products to buy, new books to read and new movies to stream based on your prior purchasing history and also your browsing history. Other companies, like Netflix, Spotify and Apple Music employ sophisticated algorithms to recommend new things to watch and listen to.

Although the algorithms that these companies use are proprietary and not open to the public, they use the same information that you use when forming concepts: experience and similarity. We form concepts by noticing the similarity among groups so that we can make predictions about things. Shopping and streaming algorithms do the same thing and are able to recommend new things that are similar to your previous experiences and interactions with the website. Concepts help companies make predictions.

These patterns can provide insights into our own behaviour and even show how our behaviour influences how companies operate. There's a great example from a psychologist at University College London named Adam Horsnby (Hornsby, Evans, Riefer, Prior & Love, 2019). Recognising that shoppers and stores both rely on categories, Hornsby and colleagues showed that people's shopping behaviours are shaped by (and also help to shape) the way grocery stores are organised. The way they did this was very creative. They collected the shopping receipts from millions and millions of shoppers at a supermarket. They then applied a computational model to look for patterns in the items on the lists. That is, they analysed the co-occurrences of items on the receipts. Milk and cereal. Beans and rice. Bacon and eggs. Pasta, tomato sauce and cheese. They were then able to extract higher-order concepts from that. These concepts tended to be organised around shoppers' goals and patterns, ranging from specific meals (e.g., 'stir fry' and 'summer salad') to general themes (e.g., 'cooking from scratch' or 'food to eat now'). The authors showed that people shop according themes and also that stores work to organise themselves around those themes, which then in turn reinforces the same theme-based shopping behaviour.

When we shop, we provide information to companies to help them organise their stores to make it easier for us to shop. Keep that in mind when you're doing your own shopping. If you buy vegan foods, for example, along with a specific brand of soap, it tells the store something about you and that becomes part of how they plan, which then becomes part of how you plan. We're each observing the other, forming and modifying concepts, and adjusting our behaviours to make predictions.

I think I've covered a good deal about why we form concepts, how they work, why they work, and how we use them to think. Let's get into a more in-depth discussion of some of the fundamental theories about how these concepts are represented in the mind. A concept is an abstraction. It does not represent all the possible features and details of each and every category member, and so different theories make different assumptions about how much information is abstracted and stored and how much is then discarded.

Theories of Conceptual Representation

Why do people seem to rely on mental categories for thinking? And why do we, as humans, seem to have the concepts and categories that we do? One possibility is that humans respond adaptively to categorise, classify and form concepts that reflect the natural structure of the world. Or as Plato had suggested, 'carve nature at the joints'. By this account, we'd argue that the reason there is structure in the mind is that there is structure in the world. I have a fairly good concept of dogs and cats because there is a natural distinction between the two groups. Cats and dogs are both members of the larger 'pets' category and may even live together but they are not the same. Cats and dogs are likely to fall into different categories whether we have nameable concepts for them (e.g. the words cats and dogs) or not. Animals, mountains, rivers, plants, rocks and rain are all structured by the natural environment. The boundaries are there in the world for us to acquire and so we go about acquiring them, learning them and naming them.

But that's not the only kind of concept. And that's not the only way to learn about categories and classes. Another possibility is that humans form concepts to help achieve goals, as the research on shopping behaviour suggests. Or we group things together and form a concept to help solve a problem. These kinds of categories might not reflect any particular natural structure. Things might be categories because we behave the same way towards them, even if they don't look the same. Because we seem to have different kinds of concepts, and because there are different kinds of categories out there in the world, psychologists have explored conceptual representation in different ways.

Like our study of memory, which includes short-term memory, semantic memory, episodic memory and procedural memory, there are a lot of possible ways to divide conceptual space. I am going to focus on four different but not mutually exclusive theories that have had an influence on cognitive science. None of these theories is the 'right' one and they overlap in some ways also. Each theory describes some of our conceptual experience pretty well, but each also has limitations too. That's the nature of any scientific theory of the mind. We're going to get some things right but will also miss some other things.

The first of the four theories is sometimes called the 'classical view' of concepts. This is an approach that is closely tied to Philosophy. This view emphasises featural rules that delineate the category and thus define the concept as a class membership that is bounded by the rules. The second of these theories emphasises the similarity relations within and between categories and also the organisation of concepts in semantic memory. I'll refer to it as the 'hierarchical view' because this theory includes the theories we discussed earlier in the memory chapter, like the hierarchal model of Collins and Quillian. The third of the four theories of concepts was developed in the 1970s as an alternative to the classical approach and is sometimes called the 'probabilistic view'. Like the hierarchical view, this theory also emphasises the importance of similarity within and between categories but does not rely on a strictly definitional approach like the classical view does. In this approach, the mental representation is an abstraction that summarises the typical features of all the members of the category. Finally, some psychologists have argued for the role of knowledge and naïve theories about the world that cannot be explained by any of the three theories above. This theory is sometimes known as the 'theory view' of concepts.

These approaches have been instantiated in several different ways, and each makes a number of core claims about how much unique information about an individual object is retained in the conceptual representation and how much unique information is lost in favour of abstracted, category-level information.

The classical view

The classical view is often described as 'classical' because this is the way that theorists understood conceptual representation from the classical, Western philosophical tradition. In addition, we can think of this view as emphasising categories as strict classes. There are two core assumptions in the most rigidly defined version of this theory. First, central to this theory is the idea of necessary and sufficient conditions as qualifiers for category membership. Second is the claim that categorisation is absolute, and all members of a class are of equal standing. This view probably seems too rigid to be realistic, but it did seem to be the underlying theoretical framework that guided work on concepts and categories for most of the early era of psychology, and in many ways is still present in the way we seek to define things, ideas and even people.

Consider a square. A square is defined as a shape with four equal sides and four right angles. That's the best definition I can come up with. As long as the shape has these features, we classify it as a square, and indeed the attributes are generally enough to allow the shape to be called a square. Any object with those features is member of the group. And that's all it takes. Colour, size, texture and substance don't really matter for classification. All that matters are the key geometry-defined attributes. That is, each of these attributes is necessary and, together, they are sufficient for category membership. The definition of the square can be said to consist of these jointly necessary and sufficient conditions. In addition, once a shape can be said to possess these features and can be deemed a member of the square category, it is difficult to imagine how anything else would increase or reduce the validity of that classification. That is, given that four equal sides and four right angles is enough to be a square, it guarantees an equality of being a square. All squares are equally good members of the square category. There are no good squares or bad squares. Just squares.

Try to imagine other examples of categories with this definitional structure: the category of even numbers includes all numbers that are divisible by two, the category of US quarters includes all 25¢ coins of certain size and shape produced by the government. Unfortunately, the definitional account starts to break down beyond these basic examples. For example, if you were to make a quick drawing of a square on a sheet of paper, right now, you would draw something that looks like a square. You might be happy to call it a square. If you showed your drawing to another person and asked, 'What is this?' they would probably call it a square. But is it? If you took a ruler and measured each side, you might find that it does not have *exactly* four equal sides. They would be close, but not exactly the same. So, by definition, it should not be included in the square category. And yet you would probably still call it a square. And although we still agree that there is a clear definition for the concept of square, if you're willing to call a poorly drawn square a square then it means you're not really using that definition to make a classification. The definition is correct, but too abstract to be used.

Another problem is that even if we all agree on a definition, some examples that follow the definition might seem better category members than other examples. A typicality effect occurs when people rate some category exemplars as being better or more typical category members than others. A simple example would be to consider the dog category. Medium-sized, common dogs, such as a Labrador, a retriever, or a German shepherd, might be viewed as more typical and as better examples of the category. Smaller, hairless dogs or very large dogs, while still being every bit as good a dog as the Lab might be viewed as less typical. It is easy to imagine the typical examples for common categories: red apples, a four-door saloon, a 12-ounce coffee mug with a handle, an iPhone. The typical items seem to come to mind almost automatically when asked to describe members of a concept. Typicality effects are even observed for categories with a strict definition. We know that even numbers have a definition (numbers divisible by 2) so all even numbers should be the same. But when asked to rate numbers on whether or not they are good examples of the even/odd categories people will rate '2' and '4' as better examples of even numbers compared to numbers like '34' and '106'. This is troubling for a definitional account of concepts because it suggests that even when all members of a category should be equally good members, people are still displaying an effect of typicality.

The typicality effect was investigated systematically by Eleanor Rosch in the 1970s and her work changed the way psychologists thought about categories and concepts (Rosch & Mervis, 1975). Until Rosch's research, the classical/definitional account – even with all its potential problems – was still the best theory available. Why was her work so influential? Well, for one thing, she asked people about their concepts. But rather than ask them to define a concept, she asked people to describe examples of what belongs in a category. For example, in one study, she asked research subjects to list all the features of common categories (such as tools, furniture, vehicles, etc.). She found that some items had many features in common with other members of the category (typical items) while other items had fewer common features and more distinctive features (atypical items). These highly typical examples were also the first ones to come to mind when subjects were asked to list category members. These highly typical examples were rated higher in typicality. In other words, these highly typical examples seemed to have a privileged status. This is a problem for a classical/definitional view of concepts because a strictly definitional account predicts that these typical exemplars should not receive any behavioural privilege. Yet Rosch showed that highly typical exemplars were classified and named more quickly. If people show a preference for typical exemplars, then they may not be relying on a definition or a set of necessary and sufficient conditions.

Rosch suggested instead that we rely on family resemblance when learning about categories. With family resemblance, members of a category resemble each other, but do not share any single defining characteristic. Imagine a large family that gathers on special occasions. It might be obvious that many of the family members look similar. Perhaps many have a certain hair colour or the same kind of eyes. But it is very unlikely that there would be one trait that could identify the family members perfectly. We can imagine many categories like this: cats, carrots, candy and Cadillacs. Each member will be similar to many other members, but maybe not similar to all other members. Rather than a definition, Rosch's work suggest that we represent the cluster of features as the central concept. The more features an example has in common with the cluster of features, the stronger the family resemblance.

Basic level concepts

Rosch suggested that we tend to form hierarchical concepts about family resemblance structure that at some levels seem to maximise family resemblance and typicality. Look at the example in Figure 9.2 (page 220) and what is the first thought or word that comes to mind? You probably thought 'dog'. But you probably did not initially think 'mammal' or 'animal' or 'German Shepherd'. At more abstract levels, what we'll call superordinate level, the concepts and category members do not overlap very much in terms of features and attributes. Between-category similarity is relatively low: animals do not usually look or act very much like other high-level concepts (e.g. plants). But at the same time, within-category similarity is also fairly low: there are a wide variety of animals and not all look or act alike. Dogs, centipedes, and bald eagles are pretty dissimilar, but are all members of the animal category. Because the similarity and feature overlap are low all around, similarity itself is not a particularly useful cue in terms of predicting category membership. At the bottom, most specific level, known as the subordinate level, there is a high within-category similarity (German Shepherds look like other German Shepherds) and also a fairly high between-category similarity (German Shepherds look a fair amount like Labradors). Again, because featural similarity is fairly high, it is not a very reliable predictor of category membership. Because similarity is too high to be useful at the most specific but not high enough to be useful at the more generic levels, we tend to classify and think about things somewhere in between. Rosch referred to this middle level, the level we use most often when we identify and think, as the basic level.

Figure 9.2: What's the first thought or word that comes to mind when you see this picture?

The basic level is a special case. Within-category similarity is high, but between-category similarity is low. Although members of the dog category tend to look a lot like other dogs, there is not nearly as much overlap between members of the dog category and other animal categories, like cats and lizards. The basic level category is the level of category abstraction that maximises within-category similarity while simultaneously minimising the between-category similarity. Because of this, similarity and/or features are a reliable cue to category membership. Dogs have a dog shape and are generally similar to other dogs and do not have as much similarity overlap with other categories. The same is true for trees, cars, tables, hammers, mugs, etc. Basic level categories are special in other ways too. Eleanor Rosch and colleagues observed that basic level categories are the most abstract level at which the objects from a category tend to have the same shape, the same motor movements, and tend to share parts. Because the contrasting categories can be easily compared, and similarity is a predictive cue, basic categories also show a naming advantage as we saw above with the picture of the dog. Basic level categories are learned earlier by children, and they are listed first when subjects are asked to list members of a superordinate category. In general, the work of Rosch and many others has shown that although objects are classifiable at many different levels, people seem to operate on and to think about objects at the basic level.

Of course, not all things are classified at the basic level all the time. People with a lot of experience sometimes classify things instinctively at a subordinate level. As an example, imagine that you were a dog breeder or show dog handler. As in the previous example, you were shown a picture of a German Shepherd dog. Unlike a novice, who would be expected to respond with 'dog', the expert might be respond instinctively with 'German Shepherd'. If you spend a lot of time working with and thinking about fine-grained distinctions and very specific classifications, you just carry that expertise with you. Subordinate level classification just becomes a habit.

Probabilistic views

I suggested that the classical view, despite its long history and the intuitive appeal of definitions, falls short in several ways. So how do we represent a family resemblance as a concept in the mind? One possibility that comes directly out of Rosch's work is the idea that category membership is probabilistic, and that classification is done by comparing things to a collection of typical features. Rather than being based on a set of necessary and sufficient conditions and residing within a strict hierarchy, a concept is thought to represent a category of things that are grouped together with shared features and overlapping similarity. In this account, referred to as the probabilistic view, category membership is not definite. There is no definition.

For example, consider common categories like the kinds we've been discussing: dogs, cats, coffee mugs, fruits, etc. Instead of a definition for each, think about the characteristic features. Think about the features that members of these categories usually have. Dogs *usually* have tails, they usually bark, have four legs, they usually have fur, etc. Some dogs have most of the characteristic features: they are medium in size, have four legs, have a tail and bark. If one of these features is missing, it may reduce the visual typicality of that dog, but it doesn't disqualify them from being members of the dog category. You may have seen a dog with a missing leg. But even if that dog does not have the characteristic feature of 'four legs', we don't consider it to be any less of a dog. In the probabilistic view, the typical examples are recognised more quickly because they share more features with other category members. In a sense, the typical category member is closer to the centre of the category. And an analogous effect might be observed with the exceptional category members. A very atypical member of a category (like bats as atypical mammals or even as atypical birds if your category for 'bird' is based on observable features) is an outlier. It really is the case that the bat is a lousy member of the mammal category. It looks like a bird, acts like a bird, and cannot see very well. A probabilistic categorisation system would assume that bats will be sometimes be misclassified and will present people with some difficulty. It is plausible that our own inability to classify them readily corresponds to the fact that bats are often feared. Perhaps one reason many people fear bats is because they do not fit into a simple, basic category very easily.

But how is this graded typicality structure, which is inherent in the categories, represented in the mind as a concept? There are two opposing accounts: prototype theory and exemplar theory. Prototype theory assumes that a category of things is represented in the mind as a prototype. This prototype is assumed to be a summary representation of the category. This can be either an average of all category members so far experienced, a list of frequently occurring features or even an ideal. According to this view, objects are classified by comparing them to the prototype. There's an advantage to this kind of representation. It's abstract. It's a collection of features, and it's optimised for thinking and behaving conceptually. There may not be one particular example of a prototypical police officer, but that abstraction would have all the features most commonly seen in police officers. Police officers who have a lot of those features would be close to this prototype and therefore classified quickly and easily. The prototype is quick and reliable, but not perfect. Like our memory in general, it's a useful abstraction but not a perfect representation.

An alternative to the prototype account is exemplar theory. Exemplar theory assumes that a category is represented by many stored memory traces, referred to as exemplars. There is no high-level abstraction as in prototype theory. Instead, the similarity among memory traces for the individual things allows us to treat them as members of the same category. Rather than classify an animal as a 'dog' because of its similarity with an abstraction, we classify it as such because it is similar to many things that we have already classified as a dog. This is an approach that has strong appeal because it eliminates the need for an abstraction process during acquisition. Because decisions are based on similarity to individual items stored in memory, exemplar theory makes many of the same predictions that prototype theory makes. Trying to tease apart the two theories is possible (and important) but not for the faint of heart. It is, more or less, what I've been doing in my research lab for the past twenty years and I'm still not sure which model or theory is the better description of human thinking.

The theory view

The classical and the probabilistic views are very much concerned with how new concepts are learned and represented, but a common criticism is that much of the research that has supported this theory has relied on artificial concepts and categories, defined in lab settings. That's true. I've carried out these studies myself. Asking people to learn to classify shapes with complex rules and/or relationships gets at the perceptual side of concepts but this kind of cognitive psychological research misses a lot of the complexity of the world and how we understand it. Dogs are more than a collection of dog features; they occupy a role in people's homes, they act and behave in certain ways, they have a cultural context. None of the theories I've described so far have had much to say about this. So, an alternative to the rule and feature theories is often called the theory view. It's also called the 'theory theory' but that's a bit much.

According to this view, concepts and categories are learned in the context of pre-existing knowledge and our own theories about the world.

223

When you learn about new things, like a new game, a new food or a new device, you are not just classifying those new objects. You are activating prior knowledge and that prior knowledge helps you to understand things about what you are classifying. The pre-existing knowledge helps to activate and prioritise features. For example, in recent years, e-bikes have become very popular. These bikes are not related to scooters or motorcycles, but are essentially bicycles with electric motors to assist, but not replace, pedalling power. Classifying them this way, as bicycles, means that you already have some theory about how they will be used and how they will operate. Where the brakes will be, how the pedals will work, and where you will be able to ride them. Your knowledge helps you know which features matter in understanding them and which do not. This puts the theory view well ahead of other theories that rely primarily on similarity to understand conceptual behaviour.

This view also suggests that attributes and features may be correlated. For example, with respect to the category of birds, many common features, such as 'has wings' and 'flies', co-occur very often. We understand these correlations as being meaningful. We understand why the features are correlated. Having wings is more than just a feature. Having wings is what allows the bird to fly. Furthermore, because the theory view relies on our existing knowledge about objects and concepts, rather than just similarity, it may be able to account for some curious findings in which people often seem to ignore their own similarity judgements. One of my favourite examples is from a study by Lance Rips (Rips, 1989). Subjects were asked to consider a 3-inch round object and two possible comparison categories, that of a quarter and a pizza. One group of subjects was asked to rate the similarity between the 3-inch round object and either the quarter or the pizza. Not surprisingly, subjects rated the 3-inch round object to be more similar to a quarter because it is closer in size. However, when subjects were asked to indicate which category they thought the 3-inch object belonged in, they overwhelmingly chose pizza. There are two reasons for this. The first is that pizza has much greater variability than quarter. Although three inches is a very small pizza, it could plausibly be a member of that category. The quarter category has very little variability and also has some very specific characteristics. Quarters must be made of quarter material and must be minted by the appropriate government authorities.

They have heads and tails. In short, although they may be more similar in size to the 3-inch round object, people are unable to classify the 3-inch round object as a member of the quarter category because they know what it means to be a quarter. Only quarters can be quarters and only subjects with that knowledge would be able to make that judgement. As with the case of correlated attributes, a prototype or an exemplar model, both of which emphasise similarity, would not handle this result easily. A machine classifier that lacked the requisite background knowledge would not be able to handle this result either. But the theory view does.

Schema theory

In all of our discussions about memory, knowledge structures and concepts, I emphasised how knowledge is stored and represented. Concepts are summary representations. But we use our concepts to think and behave. So, we're also concerned with how concepts are involved in the thinking process. How do we use concepts to carry out thinking behaviours, such as problem-solving and hypothesis testing? One theory that addresses this interaction is schema theory. A schema is a general-purpose knowledge structure or concept that encodes information and stores information about common events and situations. This representation is used for understanding events and situations. A schema is a concept in action.

Think about what happens when you go to a market (a farmer's market or a city market). Assuming that you, as the shopper, have been to farmer's markets or city markets before, you have encoded information about each event, you have stored episodic memories, semantic memories, and have retrieved and used these memories during the event. The schema is the conceptual framework that allows these memory representations to generate expectations. When you arrive at the market, you expect to see several produce vendors, you expect someone to be selling fresh cut flowers, and you expect the majority of the transactions will be conducted with cash rather than credit. You don't even need to see the cut flower vendor at the farmer's market; you may expect him to be there, and if asked later, you might respond affirmatively even if you did not actually see him. The schema is a concept that helps us fill in the background.

Sometimes, however, an activated schema will cause a person to miss features that are inconsistent with that schema. Kind of like a false memory of the kind I wrote about in Chapter 6 and Chapter 8. This general idea was demonstrated many years ago by the pioneering work of Bransford and Johnson (1973). In one study, they showed that people often miss key features in a text if those features do not fit with a given schema. Subjects were asked to read the following paragraph, which was given the title, 'Watching a peace march from the 40th floor'.

> *The view was breathtaking. From the window, one could see the crowd below. Everything looked extremely small from such a distance, but the colorful costumes could still be seen. Everyone seemed to be moving in one direction in an orderly fashion and there seemed to be little children as well as adults. The landing was gentle, and luckily the atmosphere was such that no special suits had to be worn. At first there was a great deal of activity. Later, when the speeches started, the crowd quieted down. The man with the television camera took many shots of the setting and the crowd. Everyone was very friendly and seemed glad when the music started.*

It is a straightforward paragraph, and it probably conforms to our schema of being in a city, watching some sort of demonstration or parade or civic action. We see these on the news, on social media, and on television. As a result, we can probably imagine or fill in details that may or may not be actually present in the text. If we use a concept of 'peace march' to fill in details, we could miss some of the actual details that are not consistent with the schema concept. Crucially, we might miss the sentence *'The landing was gentle and luckily the atmosphere was such that no special suits had to be worn'.* This sentence has nothing to do with watching a peace march. It does not conform to the activated schema for an urban environment or a demonstration. As a result, when subjects were asked to answer questions about the paragraph they had read, they often did not remember this sentence. We ignore details and features that do not fit with our schema.

But other research suggests that details that may not be consistent with the schema are still encoded but may not become part of the primary representation unless a new schema is introduced. The details need to find a concept to stick to. Consider a classic study by Anderson and Pichert (Anderson & Pichert, 1978). Subjects in the experiment were asked to read a passage about two boys walking through a house. In addition, they were given a context to the paragraph before reading it. One group of subjects was told to imagine that they were a house burglar (burglar schema). Another group of subjects was asked to consider the passage from the perspective of a potential homebuyer (real estate schema). These schemas orient attention in different ways and if subjects are using a schema to fill in the background, they might miss details that do not fit with that schema.

They were then asked to remember details about the following passage:

> The two boys ran until they came to the driveway. 'See, I told you today was good for skipping school,' said Mark. 'Mom is never home on Thursday,' he added. Tall hedges hid the house from the road, so the pair strolled across the finely landscaped yard. 'I never knew your place was so big,' said Pete. 'Yeah, but it's nicer now than it used to be since Dad had the new stone siding put on and added the fireplace.'

> There were front and back doors and a side door which led to the garage which was empty except for three parked 10-speed bikes. They went to the side door, Mark explaining that it was always open in case his younger sisters got home earlier than their mother.

> Pete wanted to see the house, so Mark started with the living room. It, like the rest of the downstairs, was newly painted. Mark turned on the stereo, the noise of which worried Pete. 'Don't worry, the nearest house is a quarter of a mile away,' Mark shouted. Pete felt more comfortable observing that no houses could be seen in any direction beyond the huge yard.

> *The dining room, with all the china, silver and cut glass, was no place to play so the boys moved into the kitchen where they made sandwiches. Mark said they wouldn't go to the basement because it had been damp and musty ever since the new plumbing had been installed.*

> *'This is where my Dad keeps his famous paintings and his coin collection,' Mark said as they peered into the den. Mark bragged that he could get spending money whenever he needed since he'd discovered that his Dad kept a lot in the desk drawer.*

> *There were three upstairs bedrooms. Mark showed Pete his mother's closet which was filled with furs and the locked box which held her jewels. His sisters' room was uninteresting except for the colour TV, which Mark carried to his room. Mark bragged that the bathroom in the hall was his since one had been added to his sisters' room for their use. The big highlight in his room, though, was a leak in the ceiling where the old roof had finally rotted.*

Not surprisingly, people recalled more information that was consistent with the context of the schema they were given. If they were told to read the paragraph with the context of a burglar in mind, they remembered things about famous paintings and coin collections, a closet full of furs, and the side door that was always open. If they were told to read this from the context of a prospective homebuyer, they might think about the bathroom addition, the leak in the ceiling and a newly painted downstairs. These are schema-consistent details. However, the researchers then asked their participants to consider the paragraph that they had already read from the perspective of the other context. People who read it from the perspective of a burglar were then asked to reconsider (but not reread) it from the perspective of a homebuyer and vice versa. People recalled additional detail when asked to re-inspect their memory from the alternative context. The implication

228

seems to be that the information was encoded and processed but because it did not fit a schema initially it was not recalled as well. When given a reorganising context or reorganising framework, new details were recalled. Our memories and concepts are quite flexible.

Summary

Thinking – solving problems, arriving at conclusions, and making decisions – relies on well-structured mental representations. Concepts allow us to predict things, to infer missing features and to draw conclusions. If we perceive something that fits into an existing concept or category, we have access to most of the important things we know about the objects by virtue of the concepts. Once an object is classified as a member of a category, the object can inherit or take on properties that are associated with the many other objects in the same category. Concepts are the result of organised memory and, as such, concepts allow memories to be used effectively to guide behaviour. The study of concepts offers a way to understand how knowledge and memory are optimised for adaptive thinking. Concepts allow memories and knowledge to be used efficiently and effectively at the service of other kinds of thinking.

CHAPTER 10

Language and Thinking

Thinking is the process of using mental representations to interact with the environment and to act upon the world. But thinking is more than just acting. It's planning. It's deciding. It's taking the time to consider alternatives. It's taking what we know, have seen and have heard and then acting. We have evolved neurocognitive systems to help with this goal. We rely on our perceptual systems and attention to take in information from the outside world. We rely on memories and concepts to represent what we see and hear and what we have seen and heard. But we also need to evaluate what is in our memory. We need to inspect and manipulate the contents of perception and memory. For that, we've evolved a language system. The manipulation of representations is a function of our language. Natural language gives us the power to label things, to refer to thoughts and memory, and to communicate these thoughts with other people and other minds. Human language is the engine of thought.

We can certainly learn things without language. We can select behaviours and respond to stimuli without using language. But it's nearly impossible to think without language. Try to think about something commonplace or something that happened recently. For example, try to remember what you ate for dinner last Thursday. As you try to remember this, be aware of exactly *how* you try to remember it. Pay attention to what goes on in your mind while you are remembering. Try to do this right now, without distraction, and then come back to reading this chapter.

What did you notice? First, did you actually remember what you ate? If so, how did you do it? Or how did you fail to remember? What was the actual form of the memory? How did you go about probing your memory for the information? You probably thought something like this: 'OK, what did I eat on Thursday? That was three days ago. I think I had rice and some of that spicy vegetable stew that was left over from the day before.' Or maybe you thought: 'I was working late that evening and I'm not really sure if I did have dinner.' Whatever you thought, whatever you remembered, the process probably involved some kind of inner monologue or narration. You probably asked yourself the question with your inner voice (which is part of your working memory system). You also probably tried to answer the question using language. Even if you did not carry out an extensive conversation with yourself, the thinking and recollecting that you did as you considered different memories involved your language.

In other words, your memory retrieval was guided by language. And your memories themselves also guided your inner dialogue. Each time you considered a possible memory, you may have assessed its accuracy via language. If you had to report the results of this memory search to someone else, you would absolutely need to use your language. Just try to think of something without using any language at all. It's possible, but it is not easy. Language and thinking are closely intertwined. I'm not even sure it's possible to draw a line between the two. We need language to think.

Language and Communication

If we want to understand how language is used in thinking, we should begin with understanding what language is. The study of language, or the field of linguistics, is broad and covers everything from how people use language to communicate to the formal structure of language itself. The narrower field of psycholinguistics is concerned with the cognitive psychological mechanism of language acquisition and use. I'll draw from both here, as well as more broadly from cognitive science. Let's discuss what language is, how it is used in thinking, and how the language we speak affects how we think about things.

The psychology of language as communication is a great place to begin because language seems to be a uniquely human behaviour. Our use of language in thinking arose from our much earlier and more primitive use of language as communication and as a way to plan interactions with other humans and with the environment. Although language may be unique to humans, most (or all) other animals communicate with each other. Even plants communicate with each other. Honeybees, for example, rely on a system of dances and wiggles to communicate the location of nectar to other bees in the colony. You can see examples on YouTube. Honeybees collect nectar from flowers. If you're a bee, and you come across a great new source of nectar, you need to tell the other bees in your colony, because you'll all need that nectar to produce honey for the colony. But how can you tell all the other bees if you can't really talk? When a honeybee returns from a nectar location, it performs a dance that corresponds exactly to the direction it flew and how long it was flying. The dance includes a waggle movement that corresponds to flying time and an angle that bisects its dance that corresponds to the angle of flight. These two coordinates, or bits of information, are all the other bees need. Fly in that direction, with the sun at a specific angle, for about 90 seconds and there's the nectar. The bee is communicating something to other bees that is necessary for their collective survival. But the bee is not thinking. The bee has little choice in terms of whether or not to do the dance. It's not deciding anything. It's doing what it does by nature. The bee would perform this dance even if no other bees were watching. We agree that the bees are communicating and behaving but they do not seem to be thinking or using language.

Other animals have different modes of communicating. Songbirds obviously have a well-developed and highly evolved system of mating calls and warning songs. These bird songs are unique to each species and require exposure to other bird song in order to be acquired. Dogs communicate with barks, growls, yelps, posture and the wagging of tails. And anyone with a dog knows that dogs respond to human language and non-verbal cues. Even my cat, Peppermint, *sort of* responds to some verbal and non-verbal cues.

These are all sophisticated ways that animals use to communicate. But we do not consider these to be 'language' *per se*. Unlike human language, communication in these non-human species is limited and direct. The bee dances have only one function: communicating the location of food. Bird song has a set function related to mating. A bird can only learn its own song. Highly intelligent birds, like the African Grey parrot, can learn to mimic the language of humans, but they are not using human language to carry on casual conversation or to advance an agenda. Even dogs, which are capable of very complex behaviours, are really not able to use communication abilities to consider new ideas, to solve complex problems, and to tell stories. Dogs do solve problems, but not by using language.

The great apes, specifically bonobos and orangutans, are known to be able to learn complex symbol systems. The most famous of these are Kanzi, a male bonobo, and Koko, a female western lowland gorilla. Kanzi learned to communicate from observing his mother as she was being trained on a symbolic keyboard communication system. Koko, who died in 2018, also learned to communicate with humans. Unlike the symbolic keyboard that Kanzi uses, Koko the gorilla communicated via a kind of sign language. Despite the clear cognitive sophistication of these apes, the vast majority of their communication is not arbitrary and productive but consists of direct requests and responses. In other words, unlike humans, great apes do not seem to spend much time engaging in small talk. They are not using language to direct their behaviours and the behaviours of other apes in the way that humans can. They don't seem to sit around talking to each other for the sake of talking. In other words, non-human communication and 'language-like behaviour' is used primarily to engage in direct communication or as a response to external stimuli. Non-human, language-like behaviour is not tied to thinking in the way that human language is. In this way, human language is remarkable and unique.

Remarkable and unique, yes. But what is language? What makes it so unique? What is it about this particular mechanism of information processing and communication that has given humans the ability to instruct their offspring, to offload memory to the written word, to tell stories, and to tell lies? It's a remarkable system.

Early in the history of cognitive psychology, the linguist Charles Hockett described thirteen (later sixteen) characteristics of human languages (Hockett, 1960). This list of design features is a reasonable starting point. The full list is shown in the table.

Hockett's Design features for language

Feature	Description
Vocal/auditory channel	Communication involves transfer between the vocal and the auditory apparatus. Later updated to include recognition of signing as linguistically and psychologically equivalent.
Broadcast transmission/directional reception	The language signal is sent out in many directions but perceived in one direction.
Rapid fading	The verbal (or visual in the case of signed language) signal fades quickly.
Interchangeability	A speaker of a language can reproduce any message they can understand.
Total feedback	The speaker hears everything they say.
Specialisation	The vocal apparatus used in speech is *specialised* for speech production.
Semanticity	Language has meaning and semantic content.
Arbitrariness	The linguistic signal does not need to refer to a physical characteristic of the thing it's describing.
Discreteness	Language is composed of a discrete, finite set of units.
Displacement	Language can refer to things that are not immediately present.
Productivity	The finite set of units is capable of producing an infinite set of ideas.
Traditional transmission	Language is transmitted by traditional teaching, learning and observation.
Duality of patterning	A small number of meaningless units combined to produce meaning.
Prevarication	The ability to use language to lie or deceive.
Reflexiveness	Using language to talk about language. A form of metacognition.
Learnability	Language is teachable and learnable. We can learn other languages.

These are all features of human language that suggest a unique and highly evolved system designed for communication with others and also with the self (e.g., thinking). Other species' communication systems include some of these as well, but not all of them.

Let's consider a few of them in more detail. For example, language is a behaviour that has total feedback. Whatever you say or vocalise, you can also hear. You receive feedback that is directly related to what you intended to say. According to Hockett, this is necessary for human thinking. It does not take much imagination to consider how this direct feedback might have evolved into the internalisation of speech, which is necessary for many complex thinking behaviours. Language is also productive. With human language, we can express an infinite number of things and ideas. There is no limit to what we can say, what we can express, or for that matter, what we can think. But this can be achieved within a finite system. We can say things that have never been said before, yet the English language has only twenty-six letters. There are about twenty-four consonant phonemes in English and, depending on the dialect and accent, there are roughly twenty vowel phonemes. Even when allowing for every variation among different speakers and accents, this is a limited set of units. However, the combination of these units allows for almost anything to be expressed. The phonemes combine into words, phrases and sentences according to the rules of the language's grammar, to produce an extremely productive system. Contrast this with the kind of communication that non-human species engage in. Birds, bees and bonobos are very communicative, but the range of content is limited severely by instinct and design.

Another design characteristic of human language is that it is arbitrary. There does not need to be a correspondence between the sound of a word and the idea that it expresses. In English (and other languages) there is a usually small set of exceptions, words like 'smack' or 'burp' which kind of sound like what they are describing, but this is a limited subset. Spoken or signed language does not need to have a direct correspondence with the world. This is not true for all communication systems. The bee dance that I discussed earlier is an example of non-arbitrary communication. The direction of the dance indicates the direction of the nectar source relative to the hive and the duration of the waggle indicates the distance. These attributes are directly related to the environment and are constrained by the environment. This is sophisticated communication but not language as Hockett would describe it. For the most part, the sounds that we use to express an idea bear no relation to the concrete aspects of the idea. They are, in fact, mental symbols that can link together perceptual input and concepts. Human language is a discrete, arbitrary and productive system of symbols that are used to express ideas, to communicate and to engage in complex thought and action.

Language and Thinking

Language is a complex set of behaviours and it helps us to carry out even more complex behaviours. Communicative language is essentially a 'thought transmission system'. One person uses language to transmit an idea to another person. Internal language is a form of communication with our own thoughts. Language conveys thought and makes it possible.

The linguistic duality between ideas and how they are expressed is often described as a relationship between the surface structure of communication and the deep structure. Surface structure refers to the words that are used, spoken sound, phrases, word order, grammar, written letters, etc. The surface structure is what we produce when we speak and what we perceive when we hear. Deep structure, on the other hand, refers to the underlying meaning and semantics of a linguistic entity. These are the thoughts or ideas that you wish to convey via some surface structure. These are the thoughts or ideas that you try to perceive via that surface structure.

One of the challenges in terms of understanding this relationship between surface and deep structure is that very often a direct correspondence seems elusive. For example, sometimes different kinds of surface structure give rise to the same deep structure. You can say 'I'm enjoying this book' or 'This book is enjoyable' and the underlying deep structure will be approximately (though not exactly) the same despite the slight differences in surface structure. Human language is flexible enough to allow for many ways to say the same thing. The bigger problem comes when the same surface structure can refer to different deep structures. For example, you can say 'Visiting professors can be interesting'. In this case, a deep structure that follows from this statement is that when a class is taught by a visitor – a visiting professor – it is sure to be interesting because *visiting professors can be interesting*. Another deep structure that comes from exactly the same statement is that visiting a professor at their office or home can be interesting. That is a different meaning. The surrounding context will probably make it clear in a conversation, but also suggests a challenge when trying to map surface structure to deep structure. The challenge is how to resolve the ambiguity.

Ambiguity

Language is full of ambiguity and understanding how our cognitive system resolves that ambiguity is an incredible challenge. I once saw a headline from the Associated Press regarding a story about commercial potato farmers. It read 'McDonald's fries the Holy Grail for potato farmers'. It looked funny, but most of us are quickly able to understand the deep structure here. They are not frying 'the Holy Grail'. Rather, the headline writer uses the term 'Holy Grail' as a metaphor for something that is an elusive prize. In order to understand this sentence, we need to read it, construct an interpretation, decide if that interpretation is correct, activate concepts about the Holy Grail, activate the knowledge of the metaphoric use of that statement, and finally construct a new interpretation of this sentence. This usually happens in a few seconds and will happen almost immediately when dealing with spoken language – an impressive feat of cognition.

Often, when the surface structure leads to the wrong deep structure, it is referred to as a 'garden path sentence'. The garden path metaphor itself comes from the notion of taking a walk in a formal garden along a path that either leads to a dead-end or to an unexpected or surprise ending. This is more or less how a garden path sentence works. Perhaps the most well-known example is the sentence 'The horse raced past the barn fell' (Bever, 1970). When most people read this sentence, it simply does not make sense. Or rather, it makes sense right up to the word 'barn'. As soon as you read the word 'fell', your understanding of the sentence plummets. The explanation is that as we hear a sentence, we construct a mental model of the idea. If the model of the sentence does not fit with what we hear, we need to pause and construct a new model. These mental representations of sentences are constructed as we hear them. As a listener, as soon as you hear 'The horse raced' you construct a mental model of a horse that was racing. You also generate an expectation or inference that something might come after. When you hear 'past' you generate a prediction that the horse raced past a thing, which turns out to be 'the barn'. It is a complete idea and one that makes sense to anyone. When you hear the word '*fell*' it does not fit with the semantics or the syntactic structure that you created.

However, this sentence is grammatically correct, and it does have a proper interpretation. It works within a specific context. Suppose that you are going to evaluate some horses. You ask the person at the stable to race the horses to see how well they run. The horse that was raced past the house did fine, but *the horse raced past the barn fell.* In this context, the garden path sentence makes sense. It is still an ill-conceived sentence, but it is comprehensible in this case.

Linguistic inferences

Very often we have to rely on inferences, context and our own concepts to deal with ambiguity and understand the deep structure in language. The same inferential process also comes into play when interpreting the deeper meaning behind seemingly unambiguous sentences. We generate inferences to aid our understanding and these can also direct our thinking. For example, in the United States a very popular news outlet is the Fox News Network. When the network launched in the early 2000s, its original slogan was '*Fair and Balanced News*'. There is nothing wrong with wanting to be fair and balanced: that's what we expect of most news outlets. But think about this statement. What is it causing you to infer? One possible inference is that if Fox News is 'fair and balanced', then its competitors are unfair and unbalanced. Fox does not say this, but you might make that inference on your own. The slogan, like many slogans, is simple on the surface but it's designed to encourage inferences.

This reminds me of another inference and is a cue for an episodic memory. In the early 2000s, I was a postdoc at the University of Illinois and was interviewing for faculty positions in the US and Canada. Remember: I grew up in the US and attended school in the US. And nearly all my outlook was US based. One interview took place in March 2003 at the University of Western Ontario (which is where I work now). March 2003 was the month that the United States launched the '*Shock and Awe*' campaign in Iraq. This was the opening salvo to the US led military action against Saddam Hussein's government in Iraq. The campaign began on the *very day* that I left the United States and flew to Canada for my interview. The war began while I was in flight on the way to Canada. I felt awkward being outside the country and being interviewed at a Canadian institution when my own country's government had just launched an attack that was still controversial. Canada's government led by Jean Chrétien did not support the US. But this incident gave me a chance to see news coverage of this event from a non-US perspective. There was very little internet news in 2003 and no social media (which seems hard to believe). So, I watched television in the hotel room. I was struck by the language used by newscasters in Canada. In the United States, media referred to this as the 'War in Iraq'. In Canada, the newscasters were referring to it as the 'War on Iraq'. That single letter change from 'in' to 'on' makes a big difference. The word 'in' suggests the inference that the US is fighting a war against an enemy that is in Iraq. That is, 'terrorists'. It was promoted by the US as part of the larger 'War on Terror'. Canadian coverage sometimes said the 'War on Iraq' suggesting that the United States had declared war on another sovereign country. Maybe neither term was exactly correct, but how the war was being discussed by the news media probably changes how it was perceived. How we describe things, and how we talk about them, can influence how others think about them.

Metaphor and non-literal language

If language prompts inferences, and if, as Hockett suggests, it can be used to deceive, then that means that there is always more to language than is evident on the surface. We also use language to make analogies and draw metaphors. These analogies and metaphors are both examples of non-literal language and we rely on them to help in understanding things. Metaphors often involve the activation of a related concept. In the simplest form, if you know that something (A) has a certain quality, and you were told that something else (B) is analogous to A, you would use that analogy to infer things about B.

We see examples all the time. I make analogies when I lecture in my course and explain things. I'll say, 'It's just like when you . . . ' and I'm off on a tangent. I make analogies and use metaphor when I write also. In this book, I've discussed the 'computer metaphor' or the 'hydraulic metaphor' for the mind. I've referred to the 'engine' of thinking. I've discussed the 'flow of information'. It's a habit.

You probably make analogies when you to explain things to people, too. And there are a lot of examples. Most people have seen the movie *Shrek* from the early 2000s, either as the whole movie or parts of it as video clips or memes.[32] Shrek is an available source of metaphors because it's familiar and well known. In one scene, Shrek explains to Donkey why ogres are complex and difficult to understand. He says, 'ogres are like onions'. Which is an analogy in simile form (A is like B). He later explains that 'We both have layers'. As much as I hate to explain a joke, I'm going to do it anyway (and if you have not seen this scene, it might be on YouTube). When Shrek says, 'ogres are like onions', Donkey misunderstands and focuses on the surface similarity and the perceptual qualities of onions. He wonders aloud if ogres are like onions because they smell bad. Or maybe they are like onions because they make people cry. He transfers the wrong properties to Shrek, but it's funny because they are also properties of ogres. Only later does he understand the analogy that Shrek is trying to make. That ogres and onions have layers, and the outside might be different from the inside. The joke works because it lets Shrek make his deeper analogy at the same time that Donkey makes more humorous surface-level analogies.

Non-literal language is important for understanding how to think about things individually and also as a culture. The linguist George Lakoff has suggested that conceptual metaphors play a big role in how a society thinks of itself (Lakoff & Johnson, 2008). This, in turn, can affect the things we say, the things we sell, the way we present the news, and the way we discuss politics. I gave the example earlier of a 'War in Iraq' versus the 'War on Iraq'. Each expression creates a different metaphor. One is an aggressive action against a country. The other is an aggressive act taking place within a country. Lakoff argues that these conceptual metaphors constrain and influence the thinking process. He gives the example of an 'argument'. One conceptual metaphor for arguments is that an argument is like a war. If you think of arguments in this way, you might say things like 'I shot down his arguments', or 'he totally destroyed his opponent's argument'. These statements are likely to arise from a conceptual metaphor of arguments as some kind of analogy for warfare. The '_____ is a war' metaphor seems especially prevalent in the US. In fact, many US policies have made it explicit: 'The War on Drugs', 'The War on Poverty', and 'The War on Terror', were all

---

32 Memes are fascinating because they often tap into a universal or near universal reaction to something. Reaction gifs and memes from reality shows are part of the non-literal tradition but are adapted for the online age.

formally defined positions. We fight wars on disease. The novel coronavirus of 2019–2020 was an 'invisible enemy'. People needed to 'be vigilant'. We talk about people 'winning the fight against cancer'. This is not the only way to think about public health, but it seems to be one that is 'winning in the arena of ideas'. There are other examples as well. We generally think of money as a limited resource and a valuable commodity. By analogy, we often think of time in the same way. As such, many of the statements that we make about time reflect this relationship. We might say *'You're wasting my time'* or *'I need to budget my time better'* or *'this gadget is a real timesaver'*. According to Lakoff, we say the things we do because we have these underlying conceptual metaphors, and these metaphors are part of our culture. Lakoff calls it framing. These metaphors frame our understanding and encourage inference. The term 'framing' is, in and of itself, a metaphor that calls to mind a way to describe the surrounding context.

Where do these conceptual metaphors come from in the first place? Some are cultural. Others reflect a conceptual similarity between a physical thing and a psychological concept. For example, there are many conceptual metaphors that relate to the idea of happiness being 'up'. People can be said to be *upbeat* or, if they are not happy, *feeling down*, music can be *up tempo*, a *smile is up*, a *frown is down*. All of these idioms and statements come from this same metaphor. Other examples reflect the idea that consciousness is 'up'. You *wake up*, you *go down for a nap*, etc. Another common metaphor is that control is like being above something. You can be *'on top of the situation'*, you are in charge of people who *'work under you'*, the Rolling Stones recorded a popular song called *'Under My Thumb'*. These cognitive metaphors are common in English, but they are common in many other languages as well. This suggests that there is a universality to these metaphors, and a commonality among cultures between language and thought.

Lakoff's theory has been influential since it was introduced in the 1980s, but it has taken on some renewed relevance recently since the 2016 election of Donald Trump in the US and also because of the general support for populism in many other countries. Lakoff has been thinking about and writing about language and how it affects behaviour for decades and his most recent work discusses some of the ways that popular media, news media, and statements by politicians can shape how we think. Being aware of these things is important because we don't want to be misled or duped, but our minds sometimes make it easy. Using some examples from President Trump, Lakoff points out how we can be misled without realising it. And although he's using President Trump as the primary example these things can be observed in many politicians. Trump, however, made this a central part of his governing and campaigning style.

One clear example is simple repetition. President Trump repeats terms and slogans so that they will become a part of our concept. He was famous for saying/tweeting:

> We're going to win. We're going to win so much. We're going to win at trade, we're going to win at the border. We're going to win so much, you're going to be so sick and tired of winning, you're going to come to me and go, 'Please, please, we can't win any more.'

There are seven repetitions of the word 'win' in that speech, and he repeated statements like that many times since. We also heard and saw repetitions of statements like 'FAKE NEWS' and 'NO COLLUSION', etc. Lakoff argues that the simple repetition is the whole goal. Even if you don't believe the president, you are still taking in these words and concepts. And they are often amplified by people commenting and retweeting. Activation spreads. Ideas are linked.

Donald Trump is adept at controlling the conversation by framing people and ideas. He does this in at least two ways. One is with his use of nicknames. 'Crooked Hillary' for example, to refer to former presidential candidate Hillary Clinton. 'Crooked', by the way, taps into a cognitive metaphor for untruthfulness in which we think that 'TRUTH is STRAIGHT'. By calling her 'crooked' and repeating it might seem silly, but it still has the desired effect of reinforcing the concept that she is not truthful or trustworthy. Even the slogan 'Make America Great Again' is very loaded with linguistic reference, implying that it was great in the past, that it became not so great, and that President Trump's actions would make it great again in the way that it was great before.

Lakoff suggests that we should all be aware of how these things influence our thinking. We may not agree with President Trump, but according to Lakoff, these repetitions and the use of framing and metaphors will create the associations anyway. The more times you hear them, the stronger the memory becomes. And this is not just the case with President Trump. Lakoff's message applies to other leaders, politicians and media. If you're reading this in the UK, the Netherlands, India, South Africa or Brazil, these examples can be extended, and they will probably apply as well. President Trump may be an extreme example, but framing and metaphor are all around in politics, advertising and in our attempts to influence what other people think and do.

For better or worse, language influences how we think about things. It causes us to strengthen some representations and create new memories. It activates schemas and concepts. It causes us to make inferences and draw conclusions. And it makes us able to be coerced and misled. The best defence against being misled is knowing why this happens and how to recognise it.

How does Language Influence Thought?

The discussions above illustrate how language influences the way you remember things and how you think about things. Language and linguistic context have an effect on thinking. Or, as I suggested at the outset, language is how we think and how we act. Linguists have a theory for this idea, known as linguistic relativity, that suggests that our native language influences how we think and behave. This theory assumes and predicts that there will be differences among groups of people as a function of their native language. That is, thinking is *relative* to language. The strongest form of this claim is often referred to as linguistic determinism and is also sometimes referred to as the 'Sapir–Whorf hypothesis', after Edward Sapir and his student Benjamin Whorf. This strong version of the hypothesis argues that language determines thought and can even place constraints on what a person can perceive. That is, if you don't have a word for something, that implies you don't have a concept for it either. And if you don't have a concept for something you will be unable to think about it or perceive it in the same way as someone who does have a word for it.

In general, both the strong version and the weak versions of this theory are attributed to Whorf (1956), though he referred to his theory as linguistic relativity. Before he began the study of linguistics, Whorf was a chemical engineer and worked as a fire prevention engineer. That might seem like a strange way to begin studying language, but there seems to be some connection. There's an apocryphal story which suggests that his ideas and interest in linguistics arose during the time when Whorf was a fire prevention engineer and inspector. According to this story, Whorf noticed employees smoking near canisters of gasoline, even though they claimed that the canisters were labelled as empty. If it's empty, it should be safe to smoke around, right? No. An empty canister of gasoline can be very dangerous because of the fumes. The fumes are flammable. But the workers did not realise that the empty cans were not really empty. That's because the workers labelled and conceptualised them as being empty. They were linguistically empty, but not *actually* empty. They were not physically empty. Whorf began to believe that one's native language determines what you can think about, and even your ability to perceive things. This story may or may not be accurate, but it still makes an interesting point about the difference between how one describes something linguistically and what that thing actually is. In other words, 'empty' may not really be empty.

Linguistic relativity

A famous quote from Whorf reads:

> We dissect nature along lines laid down by our native language. The categories and types that we isolate from the world of phenomena we do not find there because they stare every observer in the face; on the contrary, the world is presented in a kaleidoscope flux of impressions which has to be organized by our minds – and this means largely by the linguistic systems of our minds. **We cut nature up**, organize it into concepts, and ascribe significances as we do, largely because we are parties to an agreement to organize it in this way – an agreement that holds throughout our speech community and is codified in the patterns of our language [. . .] all observers are not led by the same physical evidence to the same picture of the universe, unless their linguistic backgrounds are similar, or can in some way be calibrated. (Whorf, 1956: 213–214; my emphasis)

Whorf is challenging Plato's notion of carving nature at the joints (which was discussed in Chapter 9). Whereas Plato suggested that there is a natural way to divide the world into concepts, Whorf suggests that concepts and categories are determined by one's native language. This is often thought of as the strongest form of linguistic relativity. In this case, the theory implies that one's native language necessarily determines thinking, cognition and perception.

Have you ever heard that claim that 'Eskimo languages have hundreds of words for snow?' A claim which is sometimes followed by the claim that people who speak these languages can therefore distinguish among many more types of snow compared to English speakers. The idea behind this claim was that if you have more terms or labels for something, you can perceive more categories. Whorf made a supposition about this, which was later picked up by media outlets and newspapers as a concrete claim, and with each subsequent version the number of hypothesised Inuit words for 'snow' grew. It is pretty straightforward that this particular claim is neither true nor relevant. Whorf himself never tested or examined the claim, and most reporting of this claim made no distinction about the many different dialects spoken by indigenous northern people. There is no one 'Eskimo language'. There are several languages spoken by northern indigenous people. Inuit in Canada and Greenland speak Inuktitut; Alaskan native people speak Yupik. English, like Inuktitut and Yupik, has modifiers that allow for many descriptions of snow. Nonetheless, this is one of those persistent myths and is fairly well-known by most people.

However, this claim that language constrains or determines perception and cognition was a bold one, and in the middle of the twentieth century was very provocative. Anthropologists, psychologists and linguists began to look for and examine ways to test this idea. One of the most important challenges came from Eleanor Rosch's research. You probably remember Eleanor Rosch from the previous chapter on concepts and for her contributions on family resemblance. Rosch noted that extensive work from anthropology found regular patterns of the language used to describe colour. For example, when looking at basic colour terms, all languages seem to contain terms for dark and light. The boundaries may not always be the same, but there are some languages with only a term for warmer and lighter shades and a term for darker shades. Red is also fairly common, and languages with only three terms always have a word for black, white and red. Red is a very salient colour for humans as it is the colour of hot things and blood. As languages shifted and evolved, some languages introduced more terms.

If Whorf's claim is correct, especially the strongest version as illustrated by his 'words for snow' example, then it implies that a language with only two colour words will tend to see the world according to those two colours. They might be able to see other colours but should have difficulty seeing the distinctions between colours that have the same name. This is not an unreasonable claim. We know that perception depends in part on having some knowledge of what you are perceiving. We've discussed how perception activates concept and how concepts can affect what we see and think about. And in basic speech perception, we tend to perceive speech sounds categorically. You will have difficulty discriminating among sounds that are not part of your native language. So, this prediction is not off base.

Rosch (Heider, 1972) carried out a test like this with an indigenous group in Papua New Guinea. The Dani people have only two words to denote colours and so their language defines colours into two colour categories. One category is called *mili* and refers to cool, dark shades, such as the English colours blue, green and black. The second category is *mola*, which refers to warmer or lighter colours, such as the English colours red, yellow and white. In several experiments, Rosch asked her subjects to engage in colour learning and memory tasks with colour cards. These cards, known as 'colour chips', were taken from the Munsell colour system which is a system of describing colour on three dimensions of hue, value (lightness) and chroma (colour purity). The Munsell system has been used since the 1930s as a standardised colour language for scientists, designers and artists. The colour chips are small cards with a uniform colour on one side, usually with a matte finish. These look a lot like what you might find at a store that sells paint.

One of the tasks Rosch used was a paired associate learning task. This is when participants are asked to learn a list of things and each thing is paired with something they already know; a word that serves as a memory cue. In Rosch's task, the things to be learned were the Munsell colour chips and each colour chip was paired with a word. Some of these colour chips were what are referred to as focal colours. In other words, these colour chips were at the perceptual centre of their category. They were selected as the best example of a colour category in the prior study with English speakers. When asked to pick the 'best example', Rosch found widespread agreement for colours with the highest saturation and English speakers could remember those central examples better. The focal colour for red was the single chip that would be identified by most speakers of English as being the best example of red. Other chips might also be called red but were not identified as the centre of the category or the best example. And still other chips might be more ambiguous. They might be named red some of the time, and at other times might appear to be another colour. You can pick out focal colours yourself. If you go to select a new colour for text in your word-processing program, you can see a wide arrangement of colours, but one probably seems to stick out as the best example of red, the best example of blue, the best example of green, etc. In other words, we would all probably agree which exact shade is the best example of the colour green. This would be the focal colour for green.

In Rosch's experiment, subjects were shown a chip and taught a new name. This was done for sixteen colour word pairs. Rosch reasoned that English speakers would have no difficulty learning a paired associate for a focal colour because it would already activate the prototype for an existing colour category. They should perform less well on paired associate learning for non-focal colours because they would not have a linguistic label to hang on that colour. That is, it's easy to remember the focal colour 'red' because it looks like your image of what red should be. It's harder to remember a colour that seems to be between red and purple because it might not have a name. Speakers of the Dani language, on the other hand, should show no advantage for most of the focal colours. That is because, if linguistic determinism is operating, the so-called focal colours would not be special in any way because speakers of the Dani language do not have the same categories. As far as linguistic determinism is concerned, they should not have the same focal colours as English speakers because they have different colour categories as a result of their native language. Being shown a focal red should not activate an existing linguistic category for speakers of the Dani language, and so they should show little difference between learning the paired associates for focal colours and learning the paired associates for non-focal colours.

This is not what Rosch found, however. Speakers of the Dani language showed the same advantage for learning focal colours over non-focal colours that English speakers showed. This suggests that even though their language has only two words to denote colour categories, they can perceive the same differences in colours as English speakers can. Thus, this appears to be evidence against a strong interpretation of linguistic relativity. The Dani language was not constraining the perception of its speakers. In many ways, this should not be surprising because colour vision is carried out computationally at the biological level. Regardless of linguistically defined categories, we all still have the same visual system with a retina filled with photoreceptors that are sensitive to different wavelengths.

More recent work continued to cast doubt on the linguistic determinism theory. Barbara Malt's research looked at artefacts and manufactured objects, and the linguistic differences between English and Spanish (Malt, Sloman, Gennari, Shi & Wang, 1999). Participants in the experiment were shown many different common objects, such as bottles, containers, jugs and jars. For speakers of North American English, a 'jug' is typically used to contain liquid, is about four litres in volume, and has a handle. A 'bottle' is typically smaller, has a longer neck and no handle. A 'jar' is typically made out of glass and has a wide mouth. A 'container' is usually not made of glass, but of plastic. Containers come in round and square shapes and are usually used to contain non-liquid products. Speakers of English may vary in terms of the exact category boundaries, but most will agree on what to call a bottle, what to call a jug, etc.

Whereas speakers of North American English refer to jugs separately from jars, speakers of Spanish typically label these things with a single term. In other words, a glass bottle, a jug and a jar might all be labelled with the term 'frasco'. If linguistic determinism held true for manufactured objects, Spanish speakers should show less ability to classify them into different categories based on surface similarity. In other words, if you speak a language that has only one term for all of these objects, you should minimise attention to the individuating features and instead tend to classify them as members of the same group. However, Malt's results did not support this prediction. English-speaking and Spanish-speaking subjects did not differ much from each other when classifying these containers via overall similarity. That is, they might have the same label for all of the different objects, but when asked to sort them into groups based on similarity, they all sorted them in roughly the same way as English-speaking subjects. The linguistic label did not interfere with their ability to perceive and process surface features. In short, these results do not support the strong version of the linguistic determinism theory.

A final example of how language affects the thinking process is demonstrated in a study by Lera Boroditsky (Boroditsky, Fuhrman & McCormick, 2011). She noted that across different languages and cultures there are differences in the metaphors that people use to talk about time. This is related to Lakoff's ideas on conceptual metaphors (discussed earlier). English speakers often talk about time as if it is horizontal. That is, a horizontal metaphor would result in statements like *pushing back the deadline* or *moving a meeting forward*. Mandarin speakers, on the other hand, often talk about time as if it is on a vertical axis. That is, they may use the Mandarin equivalents of up and down to refer to the order of events, weeks and months.

It should be noted that this is not entirely uncommon in English, especially when considering time on a vertically oriented calendar. In fact, when I look at the Google calendar on my smartphone, it is arranged in a vertical axis with the beginning of the day at the top and the end of the day at the bottom. Although I still use terms like *I've been falling behind on this project*, I am also pretty used to thinking about time in the vertical dimension. We also have English vertical time metaphors, such as doing something 'at the top of the day'. Exceptions aside, these metaphors seem to be linguistically and culturally entrenched in the idioms and statements that are produced. Importantly, these differences seem closely connected to the way written language is produced and read. English reads left to right – along a horizontal – and written Chinese reads top to bottom – along a vertical.

In order to test if the conceptual metaphor and language affects subjects' ability to understand the scene, subjects were first shown a visual prime to orient them to the horizontal or vertical dimension. They were then asked to either confirm or disconfirm time-based statements (e.g. *March comes before April*). Primes, in this case, were simple diagrams that highlighted the horizontal or vertical dimensions. For example, a picture of a black ball beside a white ball with the statement *The black ball is ahead of the white ball* is a horizontal prime. A picture of a black ball above a white ball with the statement *The black ball is on top of the white ball* is a vertical prime. Boroditsky reasoned that if a prime activated a vertical metaphor, and you spoke a language that encouraged thinking about time in a vertical dimension, you should see a processing facilitation. That is, you would be faster at judging the temporal proposition. If you saw a prime that activated the vertical metaphor, but you spoke a language that encouraged thinking about time in a horizontal dimension, then you should see some cost and would be slower at judging the temporal proposition.

This is what she found in several studies. After seeing a vertically oriented prime, Mandarin speakers were faster to confirm or disconfirm temporal propositions compared to when they had seen the horizontal prime. She found the reverse effect for English speakers. This suggests that language differences may predict aspects of temporal reasoning by speakers. This finding supports linguistic determinism. Yet subsequent studies showed that this default orientation can be overridden. For example, Boroditsky trained English-speaking subjects to think about time vertically, giving them examples of vertical metaphors. In this case, after the training, the English speakers exhibited the vertical rather than the former horizontal priming effect. Although this study shows a clear impact of language on thought, it is not strong evidence for linguistic determinism because the native language does not seem to determine how time is perceived. Instead, local effects of linguistic context seem to be doing most of the work.

Language Is How We Think

Although many different species communicate with each other, only humans have developed an expansive, productive and flexible natural language. And because language provides the primary point of access to our own thoughts, language and thinking seem completely intertwined. We use our language to investigate and describe our own memories. Memory is flexible and malleable. This flexibility can occasionally be a liability as memories are not always accurate. Memories are a direct reflection of the linguistic processes used during the encoding process and the retrieval process. We also use our language to label things in the world and link percepts to concepts. Verbal labels provide an access point to our ideas.

Animals without language use memories and have concepts, though. Animals without language behave intelligently. But human language gives us a way to think beyond the present. Human language gives us a way to think about the world, ourselves and our actions. When we take the time to think and reason carefully, it's usually by way of language, rather than by instinct or gut feeling. We talk ourselves out of bad decisions. We think through the pros and cons of something by talking to ourselves. In deductive reasoning, language use must be precise in order to determine a valid argument from an invalid argument. Language use can influence how decisions are made by providing a context or frame. The same decision can be framed as beneficial or as a potential loss. Linguistic content and semantics can have a sizeable impact on the behavioural outcome of decisions.

This chapter and the chapter preceding on concepts come the closest to addressing the central theme. How do we think? We think with our concepts. And we use our natural language to do much of our thinking.

CHAPTER 11

Thinking about Cognitive Bias

Does your mood affect how you think? Do situations and context affect your ability to reason and decide? They do for me. And you probably think they do for you also. There are probably times when you feel like you're on a roll, working in the zone, feeling the flow. Maybe this is in the early morning when you feel mentally fresh. Maybe this is when you are working on something you really love to do. Maybe this is right now, while you are reading this book. In all these cases, you might find that it feels less tiring to work and the problems seem easier to solve. But there are other times when you probably feel like you just can't concentrate. Or times when your mind feels like it's just not working. Maybe this happens when you're tired, when you're thinking about the news, or distracted by your smartphone. Researchers are speculating that thinking will be affected by crises like the COVID-19 pandemic. Why? It's been so draining and stressful for many people. Even those who are not directly affected by the virus may experience negative psychological effects and costs to their ability to think and concentrate as a result of worrying about the virus, thinking about their jobs, and thinking about an uncertain future (Holmes et al., 2020).

Most psychological research suggest that all of these things do have an effect on our ability to think. One of the most exciting areas in the field of the psychology of thinking is the study of how situational contexts, motivational factors and mood state influence people's thinking. We see examples in advertising, marketing, politics and public opinion. We see examples in our ability to make judgements and decisions when we are stressed, tired, in a good mood or in a bad mood. Sometimes it's several things at once.

I live in Southern Ontario. It's an area between two of the Great Lakes (Lake Erie to the south and Lake Huron to the west). And I used to live in Buffalo, NY, which was on the eastern shore of Lake Erie. Both of these regions can be hit with some pretty bad winter weather and sometimes we get a kind of snow called 'lake effect' snow[33] in which very cold air sweeps across the Great Lakes, picks up moisture along the way, and drops in massive, 'whiteout' blizzards. These can be sudden, very intense and localised to a fairly small region. This makes it difficult or impossible to drive. When I've been caught in a whiteout, it's hard to see, hard to drive, and after I get to work or get home, I'm mentally and physically exhausted. It's not a good time to try to do a task that demands complex thinking. My mind feels tired from the stress and effort of driving in a whiteout. I certainly don't feel confident coming off of a long stressful drive and right into an important meeting or lecture.

Regardless of where you live and how you get to work or school, most of us have has some experience of a frustrating or demanding commute to work in the morning, or a very busy morning routine. Think about how those events might affect your ability to solve a problem or make an important decision afterwards. If you have had a stressful commute to work and are immediately faced with needing to make an important decision, it is not unreasonable to think that your ability to make that decision may be compromised. In fact, some research suggests that when you have experienced some cognitive fatigue, you are more likely to use fast, decision-making heuristics. Furthermore, you are more likely to use those heuristics unwisely and to fall prey to decision-making biases.

This cognitive fatigue is an effect of context. The context of having been stressed and having to concentrate leaves you with relatively fewer cognitive resources to think. This kind of context effect comes up in other ways, too. Suppose you have received a really nice message from a good friend right as you sit down to your laptop to work. This puts you in a very good mood. Your good mood has energised you and you are able to work on a work-related problem that has been bothering you for a while. The good

---

33 Whorf might claim that people living in the Great Lakes region should have many specialised words for snow and bad weather.

mood gives you stamina to keep going and you solve the problem. Not every interaction like this prompts a good mood, though. If I'm waiting on a decision about a research grant proposal, and I'm expecting the news at some time during a specific week, I simply cannot concentrate on anything else. I wait. I check email. I refresh my browser. It's infuriating but it just seems like there's not much I can do until I get the news. It's not a time to work on complex problems because I find myself being too distracted.

A few years ago, I was lecturing at university and I was awaiting news about my younger brother who had been admitted to the hospital with a critical infection. I knew that he had been admitted as an emergency earlier that day, and I was worried about him. Certain kinds of infections can be life threatening, and this was one of those kinds. I was lecturing and this was in the back of my mind. And because I was worried, I left the vibration on for my phone in my pocket. It began to pulse, and I feared that it was a call from the hospital. As the phone vibrated, I was unable to think, speak or concentrate on what I was saying. Although normally, I would not have any special difficulty ignoring the call, this was not one of those times. I was unable ignore it and I left the lecture to take the call.

It was a marketer. Not the hospital.

I still had no news about my brother (who was fine, I heard much later). I gave a very spotty lecture that day, because I was unable to keep my mind focused. But I did use that event in class to make an example about distraction, attention and cognitive fatigue, just like I am now.

In the preceding examples, the context around problem-solving, decision-making or thinking may be having an effect on our ability to behave optimally. This is something we must consider. If we're going to improve our ability to think, we have to understand how and why context affects cognition. We have touched on this topic already. We discussed the role of context in the activation of schemas. We discussed the ability of language to frame. But in this chapter, I will discuss how physiological, contextual and social factors affect many of the core thinking processes that have already been covered. I'll begin with a sometimes-controversial theory that suggests we have two modes of thinking. If you've read Daniel Kahneman's *Thinking, Fast and Slow* (Kahneman, 2011) it's all about those two modes. There's a fast mode and a slow mode. An intuitive mode and a deliberative mode. One of these, the slow mode, uses more cognitive effort and if the context compromises the amount of resources you fall back on the faster mode.

The Dual Process Theory

Psychology has a long history of proposing two complementary processes or mechanisms for behaviour. We saw this earlier in this book with short-term and long-term memory and also implicit and explicit memory. There are theories that emphasise conscious vs unconscious processes, controlled vs automatic responses and cognitive vs emotional responses. Dual Process Theory is a meta-theoretic approach that ties a lot of these ideas together. One of the reasons it's been so influential is that it offers an organising principle for human thinking ability that differentiates some complex, language-directed thinking from other, faster intuition that might also be seen in non-human animals. It's a theory that covers most aspects of human thought. It's a big theory. It's familiar. That's also why it's been a bit controversial. This theory can also seem too all-encompassing. Because it seems to account for so much, it can be difficult to falsify. It's useful, though, despite this caveat.

Let's get some of the confusion out of the way first. Dual process theory is sometimes called the 'dual systems' account. Dual process is more common. But the two components of the dual process account are usually referred to as 'systems', which might add to the confusion. Thus, the dual process account is comprised of two systems. These are usually referred to as System 1 (the faster system) and System 2 (the slower system). Think of 'systems' as a cluster of cognitive operations, neural structures and outputs. Of course, some of the things overlap. Both systems rely on memory. Both systems are liable to make mistakes. But they differ in the way they process information.

I like to remember these by reminding myself that '1' is faster like being in a race, being the first to get started and the fastest runner. The dual process account has been one of the more influential theories about the thinking process in the last twenty years (Sloman, 1996). Much of the research on how mood affects thinking or on how cognitive fatigue affects thinking can be understood within the framework of this dual process account. Let's take a closer look at each system, how it works, and what kind of thinking each is thought to influence.

System 1

System 1 has been described as an evolutionary primitive form of cognition. This means that the brain structures and cognitive processes associated with System 1 are likely to be shared across many animal species. At the lowest level, all animal species are able to make quick responses to threatening stimuli. All animals are able to generalise responses to stimuli that are similar to threatening stimuli. The same can be said for stimuli that fulfil a basic drive. An animal can respond quickly to a potential food source, a potential mate, etc. For cognitively primitive species, we do not consider this kind of behaviour to be thinking. A mouse that moves towards a food source and away from an open space that could expose it to predators is not thinking about its behaviours. Rather, it is behaving. This is a combination of innate responding and instinct and learned associations. The larger cat that is watching the mouse and seemingly waiting for the ideal time to pounce is also not thinking about its behaviours. The cat is also behaving in accordance with instinct and learned associations. Neither the cat nor the mouse has language. Neither the cat nor the mouse has the kinds of concepts that we do. Neither the cat nor the mouse has sufficient brain mass to be able to carry out what we consider to be thinking. They cannot contemplate various outcomes. They cannot consider the pros and cons of when to run and when to pounce. The cat and the mouse don't think. They just behave.

The same mechanisms that allow the mouse and the cat to make fast decisions, influence fast decision-making in cognitively more sophisticated animals like non-human primates and humans. Humans have a lot of the same kinds of instincts that other animals do. We remove our hand from a painful stimulus without thinking about it. The neural structures that drive this instinct do not need to engage higher-level cognitive processes. Subcortical structures, such as the amygdala and the limbic system regulate emotional responses to stimuli. This is how we are able to react to hunger, and to proceed cautiously when we detect the state of potential threat. This is also how we can experience anxiety in a state of uncertainty. This is our System 1.

System 1 is not a single system but rather a cluster of cognitive and behavioural subsystems and processes that operate independently and with some autonomy. For example, the instinctive behaviours which exist in all animals are part of this system. The general associative learning system that is responsible for operant and classical conditioning is also part of the system. This includes a dopaminergic reward system, whereby a positive outcome strengthens connections between neural responses and a non-positive outcome does not strengthen the associations. Most dual process theorists assume that the information processing carried out by the collection of cognitive processes that make up System 1 are largely automatic, take place outside conscious access, and are not amenable to cognitive appraisal. Only the final output of these processes is available to consciousness. System 1 cognition is also generally conducted in parallel. That is, many of the subprocesses can operate simultaneously without cost.

System 1 provides quick solutions and decisions based on what we know. In doing so, it tends to rely on information that is relatively quick, easy and less resource-intensive to access. As a result of the tendency to rely on information that is quick and easy to retrieve, we show systematic patterns in our thinking. These are often referred to as heuristics or cognitive biases. We have already discussed several of these (availability, representativeness) but let's discuss a few more. It's not a complete list, but I want to highlight the role of System 1 in understanding. I've **bolded** the terms to make them easier to see. What is common is that System 1 provides a fast answer, based on what's in memory, on what's familiar, and on what you believe. These are frequently the right answer or a sound judgement. But not always.

Beginning alphabetically (organising things alphabetically is heuristic in its own right), there is an effect called **anchoring**. This is a general heuristic or bias in which a person bases their judgement on a common reference point or a highly salient example. For example, when given the option to donate money, you might consider donating more when the options begin at £20 than when they begin at £1. System 1 provides a fast response by considering options close to the anchor. That's just easier to think through and requires less effort. Next on the list is **availability**, which involves basing a judgement on the information that is most available in memory or that is most easily accessed. System 1 bases judgements on the information that is easiest to retrieve and reduces the cognitive demands of a judgement. With respect to logic, a **belief bias** is the tendency to accept an argument as valid just because it seems true or seems believable. This happens when we reason from memory and familiarity and not by using the language-driven and more resource-intensive forms of deductive logic.

Another common bias, perhaps the most well-known, is **confirmation bias.** This is the tendency to seek information that confirms what we believe or that confirms an existing decision or judgement. It's a bias that affects just about everything. We read news websites that we already tend to agree with. We see what we want to see in current events and discount what we don't agree with. This can reduce the cognitive processing needed for the task by reducing the number of options to consider but can be a harmful bias when we don't even consider evidence that would disconfirm what we believe. I'll have a lot more to write about this bias when I write about reasoning.

Let's list a few more. **Framing** effects occur when the context around a judgement or a decision affects how that decision is made. System 1 can provide a quick judgement on the basis of information that may be easier to retrieve because it is related to the frame. The frame, as we saw in Chapter 10, is often language-based and steers the mind in one direction or another. **Recency** is the tendency to base judgements and decisions on more recent examples from memory. This is related to availability and assumes that we remember and place more weight on recent cases. System 1 will provide a decision on the basis of information that is recent to reduce the demands of recalling older examples and memories. Finally, our old standby **representativeness** is the tendency to treat an example as representative of its category. This reduces the cognitive demands of making a decision because it takes advantage of familiar concepts and our natural tendency to generalise.

This is only a partial list of biases, but what they have in common is that they show that when we have partial information and need to make a decision or a judgement, we tend to rely on these biases as a cognitive shortcut. System 1 is responsible for these fast decisions. Most of the time, these heuristics and biases provide the right answer (or an answer that's good enough) and so we do not notice that they are biases at all. But when they provide the wrong answer, we might make these errors. One way to overcome these biases and to reduce the tendency to make errors is to slow down and to deliberate in decisions and judgements. Slow, deliberative thinking is the domain of System 2.

System 2

According to researchers like Steven Sloman, Jonathan St. B.T. Evans and Keith Stanovich (Evans, 2003), System 2 is generally understood to have evolved in humans much later than System 1. Most theorists assume that System 2 is uniquely human. System 2 thinking is slower and more deliberate than System 1. System 2 is also assumed to be mediated by linguistic processes. In other words, the contents of our thoughts are able to be described via language. We use language productively and effectively to arrive at a decision using System 2. System 2 thinking is carried out in serial or sequential fashion, rather than in parallel fashion, as in System 1. That means operations take longer. That means that System 2 thinking relies on working memory and attentional systems. In other words, relative to System 1, cognition and information-processing carried out by System 2 is slower, more deliberative, and limited in capacity. However, despite these limitations, System 2 is able to carry out abstract thinking that is simply not possible in System 1. As an example, consider the two most common ways of arriving at a simple decision. When faced with the opportunity to make a purchase, you can make an impulsive decision, based on what 'feels right', or you can deliberate and consider the costs and benefits of buying versus not buying the item. The impulsive decision is likely to be driven by the processes in System 1, whereas the deliberation is made possible by the ability to hold two alternatives in working memory at the same time, to assess attributes and to think proactively and retroactively to consider the costs and benefits. This takes time. This takes cognitive effort. And this cannot be carried out in the fast, intuitive and associative System 1. This kind of thinking can be carried out only in the slower, deliberative System 2.

We use both systems, but often end up relying on the output of System 1 because it's fast and usually adaptive. It's not always the case, and clever research can expose biases. One of the strongest paradigms to show a role for two different systems in reasoning ability is what is known as a belief bias task. Belief bias, which I described earlier, is a cognitive bias related to confirmation bias in which we tend to accept logical premises with believable conclusions, even if they are not valid deductions. Jonathan Evans (Evans, 2003) has carried out a number of these experiments in which people see logical statements that were designed to create a conflict between the output of System 1 and the output of System 2. In this case, System 1 output is the result of memory retrieval and beliefs, whereas the output from System 2 is the result of a logical deduction. Memory retrieval is fast and automatic and represents a quick way to make a heuristically based response. System 2 typically handles logical reasoning.

In the belief bias task, participants were presented with different kinds of syllogisms (as a kind of logical argument) that displayed different degrees of conflict between the output of the two systems. The first kind of syllogism was a no-conflict syllogism in which the argument was both valid and believable. For example:

> **Premise:** No police dogs are vicious.
> **Premise:** Some highly trained dogs are vicious.
> **Conclusion:** Therefore, some highly trained dogs are not police dogs.

The conclusion is the only one that can be drawn from the premises. Furthermore, it is believable that there are some highly trained dogs which are not police dogs. And so, there is no conflict between people's memory and beliefs and their ability to comprehend the logical task.

Other statements were still valid, but the conclusion was not as believable.

> **Premise:** No nutritional things are inexpensive.
> **Premise:** Some vitamin tablets are inexpensive.
> **Conclusion:** Therefore, some vitamin tablets are not nutritional.

In this case, the structure is logically valid such that the conclusion is able to be drawn unambiguously from the stated premises. However, most people regard vitamins as being nutritionally beneficial. In this case, the conclusion that some vitamins are not nutritional is less believable. There is a conflict. Another conflict argument is one in which the argument is invalid and yet the conclusion is still believable. For example:

> **Premise:** No addictive things are inexpensive.
> **Premise:** Some cigarettes are inexpensive.
> **Conclusion:** Therefore, some addictive things are not cigarettes.

This can be a challenging syllogism. It is not logically valid because the conclusion is not the only one possible, but the conclusion seems reasonable because it is believable: some addictive things are not cigarettes. This syllogism presents a conflict between the believability and the validity. Finally, participants were also shown syllogisms in which there was no conflict because it was neither valid nor believable. For example:

Premise: No millionaires are hard workers.
Premise: Some rich people are hard workers.
Conclusion: Therefore, some millionaires are not rich people.

There is no conflict here because whether you try to solve this from believability and memory or from logical reasoning, the syllogism is still false.

In the experiments, participants were explicitly told to engage in a logical reasoning task and to indicate as acceptable *only* the syllogisms that were logically valid. That is, they should endorse the first two examples shown here because they are both valid. It should not matter whether they are believable or not. But participants in Evans' experiments were influenced by the believability. That is, when there was no conflict, the valid argument was accepted more often, and when there was no conflict, the invalid argument was accepted less often. For conflict cases, there was much less clarity. According to the dual process account, these participants were unable to negotiate the conflict between the memory-based solution provided by System 1 and the logical solution provided by System 2. When there is no conflict, and the systems generate the same answer, the endorsements are correct. When there is conflict, the endorsements are incorrect. And conflict, it seems, is all around.

Take the example of impulsiveness in reward seeking. A fast system (1) will seek a fast, certain reward. A slower system (2) can wait and take into account the pros and cons, the costs and benefits. Sometimes acting fast is good. Other times, it is not the optimal approach. Sometimes it is better to wait and delay gratification. We commonly call this 'the marshmallow test'.

The marshmallow test is the common name for an effect first discovered by Walter Mischel and colleagues in the 1970s (Mischel, Ebbesen & Zeiss, 1972) but the term has entered the popular lexicon and we all use it as a shorthand for delay of gratification. If you want to see a slightly exaggerated example, search on YouTube for 'marshmallow test' and you'll see a lot of versions. None of these are standardised, of course, they are stylised to show the effect. The original study was designed to investigate the phenomenon of delayed gratification in children. Children aged four to six were seated at the table and a tempting treat was placed before them (it is often a marshmallow but is just as likely to be a cookie or something similar). The children were told that they could eat the treat now or, if they could wait 15 minutes without eating it, they could get two treats. The researchers then left the room. Typically, younger children were unable to wait the 15 minutes; they ate the treat. Other children, in order to try to make themselves last the 15 minutes, covered their eyes or turned away. In some cases, children became agitated and were unable to sit still while they waited. In general, the results showed that many children could wait, but other children could not. Age was one of the primary predictors. But the researchers speculated that personality characteristics or temperament might also be playing a role. Later research discovered that the children who were most able to resist the temptation of instant gratification were likely to have higher test scores on standardised tests later in life.

The children who took part in the original study were followed up ten years later. Many of the children who were able to delay gratification were more likely to be described by their peers as competent. Later research found that subjects who were able to delay gratification were more likely to have greater density in the prefrontal cortex whereas subjects who were less likely to delay, or who had shorter delay times, had higher activation in the ventral striatum. This area is linked to addictive behaviours. The suggestion is that the 'marshmallow test' taps into a self-regulatory trait or resource that reliably predicts other measures of success later in life.

One possibility is that the struggle between eating the marshmallow right away or sooner versus delaying for a larger reward represents the conflict between System 1 thought and System 2 thought. System 2 thought is highly associated with the prefrontal cortex areas that showed greater activation in those participants who were able to delay gratification longer. The implication is that they may have earlier access to the substructures that make up System 2. These children were able to consider the costs and benefits. These children were able to delay the gratification because they could reason whereas other participants could not delay the gratification. Relatively less well-developed processes of inhibitory control would mean that the faster System 1 would initiate and carry out the behaviour and the System 2 would be unable to override it.

The marshmallow test is not typically interpreted within the context of dual systems theory, however the notion of conflict between an instinctive response and a measured response is.

Bias in Political Debate

When I teach my course on thinking at my university, we often discuss current events in the context of the course. We all read the news, follow social media, and we're often thinking about the same things. Many of the topics lend themselves naturally to the discussion of current events and in one class, after a mass shooting in the US, I asked the following question:

> How many of you think that the US is a dangerous place to visit?

About 80 per cent of the students raised their hands. Surprising? I thought so. Most students justified their answer by referring to school shootings, gun violence and problems with American police.[34] Importantly, *none of these students had ever actually encountered any gun violence in the US*. They were thinking about it because it had been in the news. They were making a judgement on the basis of the available evidence about the likelihood of violence. This is, of course, an example of the availability heuristic (Tversky & Kahneman, 1974). As described earlier in my memory chapter, people make judgements and decisions on the basis of the most relevant memories that they retrieve and that are available at the time that the assessment or judgement is made. Most of the time, this heuristic produces useful and correct evidence. But in other cases, the available evidence may not correspond exactly to evidence in the world. For example, we typically overestimate the likelihood of shark attacks, airline accidents, lottery winning and gun violence.

Another cognitive bias that seems to be influencing people's response is the representativeness heuristic. This is the general tendency to treat individuals as representative of their entire category or concept, and I've discussed this earlier as well. Suppose someone formed a stereotypical concept of American gun owners as being violent, based on what they had read or seen in the news. They might infer that each individual American is

---

34 I raised this issue with my class *before* the COVID-19 pandemic of 2020. That would probably change the answer considerably.

a violent gun owner. This would be a stereotype that could lead to a bias in how they treat people. As with availability, the representativeness heuristic arises out of the natural tendency of humans to generalise information. Most of the time, this heuristic produces useful and correct evidence. But in other cases, the representative evidence may not correspond exactly with the individual evidence.

What does this have to do with guns? I think these biases are some of the reasons people can't seem to find any common ground on hot-button issues. It's been reported widely that the US has one of the highest rates of private gun ownership in the world (Lopez, 2019) and also has a high rate of gun violence relative to other countries. We all know that *correlation does not equal causation*' but many strong correlations often do derive from or suggest a causal link. And many people think that the most reasonable thing to do would be to begin to implement legislation that restricts access to firearms. But this has not happened, and people are very passionate about the need to restrict guns. Why do Americans continue to argue about this? Many people have limited experience with guns and/or violence and have to rely on what we know from memory and from external source and as a result, we are susceptible to cognitive biases.

Let's look at things from the perspective of a gun owner. Most gun owners are responsible, knowledgeable, and careful. They own firearms for sport and for personal protection. From their perspective it might seem like the biggest problem is not guns per se but would-be criminals who use them to harm others. After all, if you as a gun owner were safe with your guns and most of your friends and family are safe, law abiding gun owners too, those examples will be the most available evidence for you to use in a decision. You would base your judgements about gun owners and gun violence on this available evidence and decide that gun owners are safe, too. As a consequence, you conclude that gun violence is not a problem of guns and their owners but must be a problem of criminals with bad intentions. Forming this generalisation is an example of the availability heuristic. It may not be entirely wrong, but it is a result of a cognitive bias.

But many people are not gun owners. These people probably still feel safe at home and the likelihood of being a victim of a personal crime that a gun could prohibit is very small. If you do not own a gun, you would not feel that your personal freedoms would be infringed by gun regulation.

When a non-gun owner generalises from their experience, they may have difficulty understanding why people would need a gun in the first place. From their perspective, it might be more sensible to focus on reducing the number of guns. Furthermore, if you don't have a gun and you don't believe you need one, you might even generalise to assume that anyone who owns firearms might be suspect or irrationally fearful. Forming this generalisation is an example of the representativeness heuristic. It may not be entirely wrong, but it is a result of a cognitive bias.

In each case, people tend to rely on cognitive biases to infer things about others and about guns. These inferences may be stifling the debate.

It is not easy to overcome a bias, because these cognitive heuristics are deeply engrained and indeed arise as a necessary function of how the mind operates. They are adaptive and useful. But occasionally we need to override a bias and rely on System 2.

I'm always reluctant to fall back on a 'both sides' argument. That's a cognitive bias in its own right. But for the vast majority of gun owners and non-gun owners, the primary source of information comes from their own experiences. Our judgements are susceptible to cognitive biases *by design*. Those on the gun control side of the debate should also try to see that nearly all gun enthusiasts are safe, law abiding people who are responsible with their guns. Seen through their eyes, the problem lies with irresponsible gun owners. What's more, the desire to place restrictions on their legally owned guns activates another cognitive bias known as the endowment effect in which people place high value on something that they already possess, and the prospect of losing this is seen as aversive because it increases the feeling of uncertainty for the future.

Those on the gun ownership side should consider the debate from the perspective of non-gun owners and consider that proposals to regulate firearms are not attempts to seize or ban guns but rather attempts to address one aspect of the problem: the sheer number of guns in the US, any of which could potentially be used for illegal purposes. Most serious proposals are not attempting to ban guns, but rather to regulate them and encourage greater responsibility in their use.

The US really does have a problem with gun violence. It's disproportionally high. Solutions to this problem must recognise the reality of the large number of guns, the perspectives of non-gun owners and the perspectives of gun owners. We're only going to do this by first recognising these cognitive biases and then attempting to overcome them in ways that search for common ground. By recognising this, and maybe stepping back just a bit, we can begin to have a more productive conversation.

In the Mood for Thinking

Did the preceding section about guns change your mood? Does the idea of being susceptible to bias put you in a bad mood? Does the news affect your mood as well? Do you ever feel better after a certain song and feel like it's easier to think? If you do, you're not alone. Mood state has been shown to affect thinking and cognition as well. Emotions and mood have core links with physiological states. And mood is complex. There are positive and negative moods, mild agitation, rage, joy, smugness, satisfaction and disappointment. There are facial expressions that go along with these. For present purposes, I will distinguish between positive mood and negative mood in general. A finer-grained distinction would be with the kind of positive mood (e.g., happy, excited, etc.) and the kind of negative mode (e.g., angry, despondent, etc.). This is a distinction of intensity versus valence.

The effects of negative mood

It has been known for a while that negative mood narrows attention focus and reduces cognitive flexibility. I want to be clear here that I'm referring to current mood state; not a mood disorder like depression which also affects thinking. What I am referring to here is simply being in a negative, depressive mood. A temporary state. Being in a negative mood state tends to correlate with focusing on details (possibly the details that put you in a bad mood). This means that you are less likely to be distracted by irrelevant stimuli. We can see this in perception. A study by Gasper and Clore (Gasper & Clore, 2002) asked subjects to make judgements about sets of pictures that were created by having three sets of simple shapes. One of these three shapes was the target, and the research subjects were asked to choose which of two other shapes was the best match. This is called the 'forced choice triad' task, and you're being forced to choose which two of three items belong together.[35] For

---

35 We saw an example earlier in the concepts chapter when I discussed the study by Rips with the 3-inch round object being matched either to a quarter or a pizza.

example, if the target shape is a triangle made of smaller triangles, one of the stimuli would match the *local* features (it might be a square configuration of small triangles) and the other stimulus would match the global features (it would be a triangular configuration of small squares). Figure 11.1 shows an example. If negative mood narrows attentional focus, then subjects in a negative mood would be more likely to make local feature matches, and choose the square made of triangles. It's like seeing the trees instead of the forest.

Global and Local Features

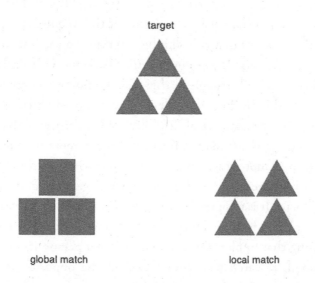

Figure 11.1: An example of the local and global feature matches from Gasper and Clore's study.

This is essentially what Gasper and Clore found. Subjects were put into a happy mood or a negative mood by asking them to write a story about a correspondingly happy or sad event. The negative mood subjects were more likely to choose matches based on the local features. Other research has uncovered similar patterns. But not all research with a negative mood has found that it narrows attention. The specific effects may depend on the actual intensity of the negative mood. In other words, being on the angry side of things may narrow your attention focus but being sad or depressed might actually broaden it. This seems intuitive. When people are feeling particularly sad, they might have trouble focusing on any one thing and feel unfocused in general. That's certainly how I feel.

The psychological research has tended to support this intuition. A cognitive experiment carried out by Gable and Harmon-Jones (Gable & Harmon-Jones, 2010) specifically manipulated their subjects' sense of sadness and found that it led to broadened attention relative to a neutral mood condition on a global–local reaction time task (similar but not exactly like the one described above). This implies that a negative, depressive mood will have a mild effect on anything that depends on cognitive flexibility. Indeed, that seems to be the case with cognitive tests like the Wisconsin card sorting task (Merriam, Thase, Haas, Keshavan & Sweeney, 1999). The Wisconsin card sorting task is a standardised assessment that requires people to learn a rule and then discard that rule, inhibit attention to the features that correlate with it, and switch to another rule. Patients with damage to the pre-frontal context will struggle in this task. Children sometimes struggle in the tasks. And individuals with depressive symptoms struggle with this task as it requires some degree of flexibility and cognitive inhibition.

Being in a negative mood can affect thinking in more than one way and, like many things, more research is needed to clarify exactly what is going on. General negative mood and/or an angry negative mood may narrow attentional focus. It is tempting to imagine an evolutionary reason for this. Perhaps if a person is in an angry mood, they want to focus on the thing that is causing that unpleasant state. But when a person is in a depressive negative mood, research has tended to find the opposite. A depressive mood seems to broaden the attentional focus, interfere with the ability to attend selectively to a single stimulus, and inhibit attention to competing stimuli. This too has an intuitive connection to general depressive thoughts. One possibility is that people who suffer from depression have difficulty inhibiting negative cognition. This may be an overall cognitive style.

Positive mood

But what about a positive mood? What about the power of positivity? I think of the positive feeling I get if I hear an upbeat song. The Beatles' *Here Comes the Sun* is a great example. When I hear that song, I think of how it feels when winter's over, the sun is out, you're in a good mood and want to spread the mood. When people are in a positive or happy mood, things seem different. A task that might seem challenging under normal circumstances might seem fairly easy when you are in a good mood. It is a common cliché to say that 'time flies when you're having fun'. Metaphorically, this relates to the notion that when we are in a happy mood, we are able to be engaged in what we are doing. We don't tend to watch the clock distractedly. Given these colloquial and popular benefits to positive mood, it is worth examining research on positive mood and thinking.

Positive mood has been associated with the enhancement of an array of cognitive skills, including creative problem-solving, recall of information, verbal fluency and task switching. Positive mood has also been associated with cognitive flexibility. A few years ago, some of my graduate students and I investigated the effects of mood on a category learning task (Nadler, Rabi & Minda, 2010). In our experiment, people were asked to learn one of the two different classification problems. One of these problems benefited from a degree of flexibility (testing a hypothesis to find a rule) and the other did not benefit because there was no rule to be found. People had to learn stimulus response associations. Another way to describe this is that some people learned a problem that needed some System 2 (the flexibility task) and others learned a task that depended on System 1 (the association task).

Before subjects learned to solve the problems, though, we manipulated their mood. We induced a positive, neutral or negative mood in our subjects, and then asked them to learn either the rule-defined or the non-rule-defined category set. To put our subjects in a positive mood, we asked them to listen to some happy music (Mozart in this case) and watch a video with a very happy laughing baby (you can probably still find this video on YouTube). We used a similar technique for the negative mood and the neutral mood except we used corresponding music and videos. We found that the subjects in the positive mood condition performed much better on the task that needed flexibility. But positive mood did not seem to have an effect on performance in the associative task. In a problem that did not benefit from flexibility, there was no benefit for being in a good mood. In other words, being in a good mood seemed to enhance the cognitive flexibility of our participants, enhance their System 2 thinking, and improve their performance, but only when the task required that flexibility.

Cognitive Resources

At the beginning of this chapter, I gave the example of driving in difficult conditions and then trying to carry out sophisticated thinking. I suggested that it might be difficult because cognitive resources would be taxed or depleted after a difficult drive. In other words, when your mind is tired and when your cognitive resources are depleted, performance on other tasks might suffer.

This idea, that cognitive resources are limited, has given rise to a now-controversial idea called 'ego depletion'. The idea of ego depletion comes from work by Roy Baumeister and colleagues (Baumeister, Bratslavsky, Muraven & Tice, 1998). This theory argues that self-regulation is a finite resource. You can use it up, just like you use up physical resources. Baumeister argues that cognitive resources and self-regulation are analogous to physical stamina. After a hard workout or a long walk, your muscles are tired. According to the ego depletion theory, your self-regulatory resources work in the same way. That is, they get tired. And if these resources get tired, your performance suffers.

Maintaining performance in a demanding cognitive task can deplete resources, and these depleted resources are known to have a detrimental effect on subsequent tasks that depend on them. Baumeister and colleagues originally used the term 'ego depletion' as an homage to Freud (Baumeister, 2014) because Freudian theory emphasised the idea of self-regulatory resources. However, they stress that their theory does not bear a theoretical resemblance to Freud's theories.

In an early study on ego depletion, Baumeister found that being asked to perform a challenging act of self-regulation can affect performance on a subsequent executive function task, suggesting that the two types of task share resources. For example, they found that when participants forced themselves to eat radishes instead of chocolates, they displayed reduced persistence on a subsequent puzzle-solving task compared to participants who did not exert self-control over eating. In other tasks, subjects were asked to watch movies that would normally elicit a strong emotional reaction. In the ego depletion manipulation, these subjects were asked to suppress any emotional reaction or distress. These subjects displayed impaired subsequent performance at solving anagrams.

A more recent study showed that participants who were asked to engage in a task that depleted their cognitive regulation resources, such as regulating their emotions, controlling their attention, or taking a working memory test, performed more poorly on subsequent tests of working memory span and inhibitory control. This suggests a correspondence between ego depletion and System 2 thinking because both of these executive functions fall under the System 2 heading. In fact, ego depletion also influences decision-making abilities. Depleted participants tend to make poorer decisions and fail to take into account decision alternatives as well as control individuals. Depleted individuals also tend to depend more heavily on heuristics and often fail to weigh all of their options carefully.

It should be noted that the ego depletion phenomenon is not the same as general fatigue. In other words, ego depletion corresponds to the condition of having one's self-regulatory resources in a state of depletion. This is fatigue limited to cognitive control. This is not the same as overall tiredness or overall fatigue. This distinction was made in a clever task that relied on sleep deprivation as a comparison to ego depletion. If ego depletion is the same as general fatigue, then ego depletion subjects should perform in the same way as sleep-deprived subjects (Vohs, Glass, Maddox & Markman, 2011). However, the research does not support this conclusion. Sleep-deprived participants suffered from fatigue and did not display the ego depletion effects. These authors argue that, unlike general fatigue, ego depletion is the 'exhaustion of the inner energy that modulates unwanted responses'.

Caution and concerns

The original idea of ego depletion has been a very influential theory. And it is intuitive also. We do feel mentally exhausted after doing something that requires us to concentrate, like taking an exam or doing our taxes. But some of the research on ego depletion has come into question. Many psychologists argued that it might be an unstable effect with limited use. In fact, it may not be much of an effect at all. The reason is that many of the core effects have not been replicated. That is, some labs have been able to find evidence for ego depletion, but other labs have not been able to reproduce the effects using the same techniques and the same methods. This idea, replicability or reproducibility, is crucial for science. One of the ways that we can trust the conclusions and interpretations is to be confident that an effect was not a fluke. So, a replication of a study can verify that the experiment works the way the original authors claimed. Like a recipe, if you follow the published methods, you should be able to obtain similar results most of the time.[36]

---

36 But not all of the time. Just like there are false positives, there should be the occasional false negative.

This is what some researchers have tried, but the results are anything but clear. For example, a large, multi-lab study attempted to replicate a specific effect of ego depletion, that concentrating on crossing out letters in a text will interfere with a subsequent task that requires inhibitory control. Several psychology labs across the world all used the same task, the same materials, and tried to obtain the same results. Although the original paper found the expected effect of ego depletion (Sripada, Kessler & Jonides, 2014), an analysis of the replication attempts did not (Hagger et al., 2016). There was no effect of ego depletion overall. Does that mean the effect is not real? It's not clear, because even more recent work, with more rigorous controls and larger sample sizes, has demonstrated that ego depletion is still a robust effect and criticised the earlier replications for not being methodologically sound (Garrison, Finley & Schmeichel, 2018).

I wish I could write about ego depletion more decisively, but in my opinion, this issue is not settled. This is a good example of psychology as a science. As more data is acquired, we need to update our theories and models. Ego depletion may prove to be stable with some additional constraints or we may arrive at a better explanation of the phenomena.

The topics in this chapter covered thinking in different contexts, but, behind the scenes, these contexts reveal the possibility of two different systems being responsible for human thought. System 1 is involved in making fast, instinctive and intuitive decisions. At the same time, System 2 is involved in making slower, more deliberative decisions. Sometimes a contextual or cognitive factor can interfere with one or the other system and this can have beneficial or deleterious effects on cognition.

A major caveat to much of this is that there are now significant questions in the field around the stability of some of these findings. Many of the more dramatic demonstrations of ego depletion and the larger literature on social priming have been unable to be replicated satisfactorily. Does this mean that the theory of ego depletion is no longer viable? Maybe. But it could also mean that our understanding of the effects of context, motivation and mood on thinking are still evolving. Theories that attempt to explain an effect without a solid grounding in stable constructs can run into difficulty when they are updated and modified to explain new effects. But new data and more rigorous methodology will improve our understanding of human behaviour.

CHAPTER 12

Predicting the Future

I have covered a lot of cognitive psychology, cognitive science and neuroscience in this book so far. I hope that I've given you a good understanding of how these fields developed and why they matter. You should know something about how your visual system takes in information from the world, how the brain processes that information, how you can switch your attention from one thing to the other, how you use your memory to bring structure to the world, and how you coordinate that with language and concepts. It is an elegant system. Information flows from the outside world to your sensory and motor systems, but it only makes sense to you when it's merged with the contents of your memory. You only perceive what you have concepts for.

It is an elegant system because, for the most part, it can be passively computational. A good deal of what I have discussed is accomplished by connected networks of neurons. It is an elegant system because we can imagine building a computer to do a lot of the same things and in the same way. It is also elegant in that it seems to have evolved to allow us to achieve goals. And many of the underlying principles are seen in other species. Rats learn associations in the same fundamental way that we do. Birds remember where they stored food in the same fundamental way that we do. It is elegant. But it is primarily *information processing.*

I've described a system and an architecture that can support thinking, but I have not described thinking yet. You might find this unfortunate, because this is a book on 'How to Think' and I have only really scratched the surface of what thinking is and how to do it. But I had no choice. In order to be better at thinking, you need to know how people think. In order to know how people think, you need to know how the mind processes information. In order to know how the mind processes information, you need to know about cognitive architecture, cognitive psychology and neuroscience. We've covered these foundations. Now let's talk about thinking.

I'm going to begin this section like I always do. With a story. These stories are sometimes actual episodes, but other times they are amalgams or abstractions that combine many experiences into a single idea of an event. This is the latter. The details are true, but the events may have occurred at more than one event. But as I discussed in Chapters 6 and 8, this is an unavoidable part of how memory is organised and how we access it.

Have you ever been shopping at a farmers' market? Or if not a farmers' market, imagine shopping at a food store, a grocery, a vegetable stand or a produce vendor. In Canada and the US, we buy most of our food in grocery stores, and a lot of the food we eat comes from other parts of the world. Lemons and limes from Mexico, cucumbers from the US, tomatoes from greenhouses in Ontario. But in a lot of places, we have a lot of locally grown food in the late summer and early fall.

Remember, I live in Southern Ontario, in the Northern Hemisphere and we have long growing season by Canadian standards but short by world standards. The summer temperature can reach +30 (°Celsius) and the winter can reach −20 (°Celsius). But in late summer and fall, or the months of August–October, there is usually a big emphasis on local, traditional produce. Things like sweetcorn, berries and peaches are popular in July and then apples, winter squash and late summer tomatoes are seen in quantity in the early and late fall. Our thanksgiving holiday is the first Monday of October when this harvest is at its peak, and we all go out and buy pumpkins and other winter squash.

A few years ago, I was at one of the local markets and was really fascinated by the looks of and the variety of all the winter squashes. There are butternut squashes, acorn squashes, pumpkins and buttercup squashes to name a few. And these are also sold beside decorative, non-edible gourds that people buy for how they look. I had purchased some butternut squash and pumpkin for roasting a pie, when my eye caught the biggest and ugliest squash I had ever seen. It was larger than a soccer ball, more like a large rugby ball. [37] It was also an ugly colour. Uglier than any other edible thing. A sickly grey-green, not dark exactly and almost pale. Covered with warts. I could not figure out what this was doing here. It did not look appetising and it seemed too big to be a practical squash to roast in the oven, but also way too ugly to be a decorative squash. It looked kind of pathetic sitting by the colourful pumpkins and gourds. Who would buy this?

'What kind of squash is this?' I asked, already making some inferences about what it might be based on the features that I could see and the surrounding context. It was being sold near winter squashes, so of course it's also a squash, what else would it be?

'It's a Hubbard squash,' the seller responded.

'What is a Hubbard squash?', I replied. 'What are they good for, do you eat them or are they just for display?':

'Well, to be honest, not too many people buy them, and we don't grow a lot of them,' he answered. 'But they are supposed to be really good, and you can make a pie out of them like you would with Pumpkin.'

Now on the one hand, I'm thinking: Why should I buy this, if it's pretty much the same as a pumpkin, why bother? Like, just buy a pumpkin. But on the other hand, it's an interesting looking squash.

So, I bought it. I had no specific plan, but I was armed with some general knowledge about what it was (a winter squash) and what it might be like (it was like a pumpkin). With this general knowledge, stored in my concepts as described in Chapter 9, I decided to make some predictions. I was going to predict the future. Not an exciting future prediction, not like predicting the outcome of an election, or the course of a pandemic disease spread, or the outcome of a sports event that I bet on. This was a simple prediction, but as we'll see, it's really the same process.

My prediction was simple. When I cut into this squash that I've never seen before, I predicted that I would find yellow/orange flesh, stringy fibres and seeds. Furthermore, I predicted that if I scrubbed out the seeds, I would be able to roast the pieces of this ugly Hubbard squash and that it would

---

37 I'm asking you to think about a related concept (soccer ball or rugby ball), to use your memory to help fill in the details, by the way. This ought to be a familiar technique by now and it's an analogy like we discussed in Chapter 10. You might never have seen a squash like the one I am describing but you have probably seen soccer balls so you can use what you know to try to imagine a thing you have never seen before.

taste like a pumpkin or an acorn squash. And that if I made a pie in the future with that roasted squash, it might have the overall taste of a pumpkin pie. No surprise; the predictions turned out to be true. Consistent with its category, the Hubbard squash has yellow, stringy fibres. Consistent with its category the Hubbard squash also tastes a lot like other winter squashes. And, consistent with its category, the Hubbard squash makes a really good pie. It was great to be able to test my predictions, but the important thing is that I was even able to make these predictions at all.

Now you might be rolling your eyes a bit at my calling this 'predicting the future' but that's exactly what this is. I can rely on all the cognitive processes we've been discussing so far – perception, memory, spreading activation, concepts and language – to make clear and direct predictions about the outcomes of my action. I use that to plan decisions and decide what actions to take before I take them. It happens quickly. It happens automatically. It relies on output from System 1 and/or System 2, and it's how we are able to survive in the world. Being able to predict the future means that you can discover new things and new ideas just by thinking about them. This is a very powerful thing.

As psychologists, we usually refer to this as an inference. Concepts and memory are activated and properties, whether they are actively present or not, are also activated. Those properties that are present are confirmed and those that are not are inferred. This can be a fairly passive and associative process. Non-human animals and machines make predictions and inferences all the time, and we discussed that in Chapter 2 when we discussed behaviourism and later in Chapter 8 on memory and Chapter 9 on concepts. Stimulus generalisation is one of the things our brains evolved to do. Humans, with the capacity for creating concepts and the abilities of language, can make and evaluate inferences to plan out actions. We call this inductive reasoning, and it's a fundamental kind of thinking. It's the basis for our survival.

I want to cover a few basic ideas about induction, it's place in cognitive science, and then discuss some specific theories that focus on how inductions are guided by our concepts and categories.

Conclusions Based on Observations

Induction, or inductive reasoning, is one of the fundamental cognitive processes that humans and many non-human animals rely upon for survival.

It's critical to know what is going to happen next. Most importantly, we use the process of induction to make inferences. Inferences are predictions and conclusions based on the available or observable evidence (which, as we know, is susceptible to bias). These conclusions might be used to make a prediction about a specific event or about a broad category of things. As an example, for many years, I used to get a lot of telephone calls from marketers between the hours of 4:00pm and 7:00pm. I don't get these calls as much anymore, primarily because I no longer have an old-fashioned 'landline' telephone. I still get them, but the numbers are easier to ignore and easier to block. These were calls, sometimes a live person and other times a recording, that might have been an attempt to sell something or convince me to buy additional services.

So, why did marketers call at that time? It's simple: the hours between 4:00pm and 7:00pm would be a time when many (though not all) people are home from work or school, making and/or eating dinner, and home for the afternoon and evening. When the phone would ring at that time, I usually made an inference or prediction that the caller was just trying to sell something, so I rarely picked up the call. I based this prediction on my memory for things that had happened in the past. Because this same event had happened many times in the past, I made enough observations to draw a reasonable conclusion about who will be on the phone. Using the past to predict the future, I could make a decision not to pick up the phone and take the call. And I was not the only one making the inference. On the other side, the telemarketing company relies on their evidence to make an inductive inference that people will be at my home phone between 4:00pm and 7:00pm. We are both making inductive inferences. That is, we rely on observations made in the past to make specific predictions.

But we do more than make specific predictions. We also rely on induction to make generalisations. A generalisation is also an inductive conclusion, but rather than describing a specific prediction, as in the previous example, generalisation is a broad conclusion about a whole class or group of things. These generalisations inform the conclusions that we make, and these conclusions affect our behaviour. If you enjoy a really good espresso at a particular café several times in a row, you will probably begin to form a generalisation about that café and that will affect your expectations. It's not just that you predict that the next espresso will be good (a specific inference) but you draw a general conclusion about that café: they pull a good expresso. On the other hand, if you had a bad dinner at a restaurant, you might form a general impression of its poor quality and that would affect your predictions about future meals and would reduce the likelihood that you would want to eat there. You are using your past experience to generate a mental representation, the generalisation, that you will use to guide your behaviour.

We also form generalisations about people based on our experience with one or more individuals. In Chapter 9, in the discussion of memory and concepts, I conisidered the possibility that people form an impression, a concept, of police officers based on their own direct experience and also indirect experience. Interactions, images, news and stories all contribute to this concept. If you see and experience the police as being friendly and helpful, you will tend to enrich your concept accordingly. That concept is what enables you to make predictions and generalisations. If you notice that features and attributes correlate and co-occur, you don't need to see them all the time to know that they are present. In other words, you don't need to see evidence of the police behaving in a helpful way if your concept has already formed that generalisation. The concepts activate these properties automatically and you expect them. And of course, the same thing happens if your concept is built up of and based on negative, violent and/or aggressive images of the police. The concepts are abstractions designed to support these inferences.

We treat people and *prejudge* them, in ways that are described and dictated by concepts that have been abstracted from experience. This is not always a good thing. It's often harmful to us, harmful to others, and harmful to the public at large. It's the basis for stereotypes, prejudice, bigotry and racism. And it's not easy to avoid. Our brains are designed to observe, perceive, abstract, match and predict. These inferences and generalisations are a natural consequence of the way the mind is organised. It can be frustrating, of course, to know that we have these tendencies. But just like our earlier discussion on memory and how the errors come from the same mechanisms as the successes, this tendency to make inferences is nearly always helpful for us. It's how the mind (or minds in general) have evolved. It's necessary for survival. Most of our inferences are benign and go forward without us even noticing.

We don't notice these inferences, because we make these kinds of inferences all the time. If you call a restaurant to place an order for pick up/delivery/takeaway, you make a basic inference that the food you order will be ready for you to pick up. When the driver in front of you puts on his indicator, you make an inference that he will turn left or right. We rely on induction to make inferences about how people will behave and react to what we say, or to make inferences about how to use new ingredients when cooking dinner. Young children rely on induction when they pick up an object and learn about how its size predicts its weight. Parents make inductions when they predict how their young children might behave after a short nap or a long nap.

The list is extensive because induction is such a critical aspect to the psychology of thinking. In summary, we rely on inductive reasoning to predict, generalise, reduce uncertainty and to discover something new by thinking.

How Induction Works

Induction is central to our thinking. And as a result, philosophers and psychologists have been thinking about and studying induction for centuries. Let's look at a very brief history of induction as an area of study. This history is fascinating because it's full of paradoxes and quandaries and many of these ideas are still relevant today.

In the era of the Scottish Enlightenment, a period of intellectual activity in Scotland in the seventeenth and eighteenth centuries, the philosopher David Hume considered induction to be one of the greatest problems for philosophers to solve. Unlike deductive logic (which I discuss in the next chapter, and that many philosophers believed could be explained by formal, mathematical operations), induction seemed to Hume to defy logical description. Induction, as we have already described, is essentially the act of relying on past experiences to make inferences, predictions and conclusions about the future. It sounds pretty basic and elementary. It's how we learn. Everyone knows that we do make inferences. Everyone knows that animals do this too. Hume knew it too. So, what's the problem? Hume was concerned that trying to explain induction ends up being a circular argument. A circular argument is one in which you try to explain a concept by relying on the same concept you are trying to explain. Hume's problem is as follows: induction works because we assume that the future will resemble the past in some way. The sun rose in the east yesterday, as it did the day before, and I assume it will do so tomorrow. In order for inductions to be useful to us, we must have confidence in our judgements about the future. Hume claimed that this only works because the future has always resembled the past, *in the past*. To say that the future has always resembled the past, in the past, might strike you as obvious, and maybe confusing. But what this means is that your inductions and conclusions were probably correct in the past. You might be able to recall inductions and conclusions that you made yesterday, two weeks ago, or two months ago that turned out to be true. As a concrete example, if you were at a farmers' market yesterday, and you made an inference about what the inside of a Hubbard squash would look like, and your prediction was later confirmed, you could say that *yesterday, the future resembled the past*. So, it is our experience, based on past observation, that the future tends to resemble the past.

The problem with this, according to Hume, is that we cannot use these past inductive successes to predict future inductive successes. We simply cannot know if the future will resemble the past. It is impossible to know if your inductions will work in the future as well as they worked in the past without resorting to the circular argument of using induction. Just because your inductive inferences worked yesterday, two weeks ago, two months ago, does not guarantee that they will work now, tomorrow or two weeks from now. Induction is based on the understanding that the future will resemble the past, but we only have information about how well this has worked in the past. To make this assumption requires the acceptance of a circular premise. In essence, we are relying on induction to explain induction. Not good, according to Hume. Not good at all.

By now, your head might hurt from considering past futures, past pasts, current presents that were once past futures, and future pasts. And you would be right. It is confusing. Hume concluded that from a strictly formal standpoint, induction cannot work. Or rather, it cannot be described logically. But it *does* work. Humans do rely on induction. This is why Hume considered induction to be a problem. There is no way for it to work logically, and yet we do it all the time. We rely on induction because we need to. Hume suggested that the reason we rely on induction is that we have a 'habit' of assuming that the future will resemble the past. In a modern context, we might not use the term 'habit', but instead would argue that our cognitive system is designed to track regularities in the world and we then make conclusions and predictions on the basis of those regularities. Let's consider some the fundamental mechanisms that allow induction to work. This does not exactly solve Hume's problem. But it solves our problem of needing to understand the neurocognitive bases of induction.

Basic learning mechanisms

All cognitive systems, intelligent systems and non-human animals rely on the fundamental processes of associative learning. There is nothing controversial about this claim. The basic process of classical conditioning (which featured in Chapter 2 when discussing the psychology of behaviourism) provides a simple mechanism for how inductions might work. In classical conditioning, the organism learns an association between two stimuli that frequently co-occur. Later, in Chapter 9, when discussing concepts, I wrote about my cat, Peppermint (Pep) and how she learns the association between the sound of the can of food opening and the subsequent presentation of her favourite food. Pep has learned that the sound of the can being opened always occurs right before the food. She has also learned to form a concept that represents the universe of food-can sounds. Although behaviourists tended to talk about it as a conditioned response, it is also fair to describe this as a simple inductive inference. Pep doesn't have to consider whether or not it is reasonable that the future will resemble the past; she simply makes the inference and acts on the conditioned response. In other words, Pep makes a prediction and generates an expectation. This is the basis of the 'habit' that Hume is talking about.

In addition to understanding induction as a habit, another advantage of relating induction to basic learning theory is that we can also talk about the role of similarity and stimulus generalisation. Consider a straightforward example of operant conditioning (again from Chapter 2). Operant conditioning, somewhat different from classical conditioning, is characterised by the organism learning the connection between a stimulus and a response. Imagine a rat[38] in a Skinner box or operant chamber learning to press a

---

38 I'm using the rat example here, because that's the common example from behaviourism and also because I just hate the thought of imagining my cat in an operant conditioning chamber.

lever in response to the presentation of a coloured light. Suppose that there are red lights and blue lights and that if a red light goes on and the rat presses the lever, it receives reinforcement in the form of rat food. But if the blue light goes on and the rat presses the lever, it receives no reinforcement. Not surprisingly, the rat learns pretty quickly that it needs to press the lever only when the red light comes on. We can argue that the rat has learned to make inductive inferences about the presentation of food following various lights.

However, the rat can do more than just make a simple inference. The rat can also generalise. If you were to present this rat with a red light that was slightly different from the original red light that it was trained on, it would probably still press the lever. Its rate of pressing might decrease, however. You would also find that the rate of pressing would decrease as a function of the similarity of the current light to the original light. The more similar the new light is to the training light, the higher the rate of lever pressing. There would be a function known as a response gradient in which responding (lever presses) is related to similarity along an exponential function, a curve. Very similar lights receive a lot of presses, but pressing rate drops off steeply as similarity decreases. This decrease is known as a generalisation gradient. So pervasive is this generalisation gradient in behaviour that the pioneering psychologist Roger Shepherd referred to it as 'the universal law of stimulus generalisation' (Shepard, 1987). Shepard observed this effect in humans and animals to nearly all stimuli. He writes:

> *I tentatively suggest that because these regularities reflect universal principles of natural kinds and of probabilistic geometry, natural selection may favor their increasingly close approximation in sentient organisms wherever they evolve.*

What we see is that there is something fundamental and universal about generalising to new stimuli as a function of how similar they are to previously experienced stimuli. This has implications for understanding induction. First, it strongly suggests that Hume was right: we do have a habit to behave as if the future will always resemble the past and this tendency is seen in many organisms. Second, our tendency to base predictions about the future on similarity to past events should also obey this universal law of stimulus generalisation. Shepard effectively solved Hume's problem of induction by providing a description of the habit to generalise. If one's past experiences are very similar to the present situation, then inferences have a high likelihood of being accurate. As the similarity between the present situation and past experiences decreases, we might expect these predictions to have a lower probability of being accurate.

Goodman's 'New Riddle of Induction'

Although stimulus association and generalisation seem to explain how induction might work at the most basic level, there are still some conceptual problems with induction. According to Hume (and indirectly, Shepard), induction may be a habit, but it is still difficult to explain in logical terms without resorting to some kind of circular argument. Hume's concern was not so much with how induction worked, but rather that it seemed to be difficult to describe philosophically. Nelson Goodman, the twentieth-century philosopher, raised a very similar concern, but the example is somewhat more compelling and possibly more difficult to resolve (Goodman, 1983).

Goodman's example, what he called a 'New Riddle' of induction, is as follows. Imagine that you are an emerald examiner. You examine emeralds all day. Every emerald you have seen so far has been green. So, using your knowledge of the past and attempting to predict the future, you can say that 'All emeralds are green'. By assigning to emeralds the property of green, what we are really saying is that all emeralds that have been seen in the past are green and all emeralds that have not yet been seen are also green. We're making a generalisation. Thus, 'emeralds are green' predicts that the very next emerald you will pick up will be green. This inductive inference is made with confidence because we have seen consistent evidence that it is true. This is so straightforwardly obvious that it's hard to see what the riddle is.

But there is a problem with this. Consider an alternative property, that Goodman called grue. And that if you say that 'All emeralds are grue', it means that all the emeralds that you have seen so far are green and all emeralds that have not yet been seen are blue: green emeralds in the past, but blue emeralds from this moment forward. Yes, this sounds ridiculous, but Goodman's point is that at any given time this property of grue is true. In fact, both properties are true given the evidence of green emeralds.

Your past experience (green emeralds) is identical for both properties. Goodman's riddle is that both of these properties, *green* and *grue*, can be simultaneously true, given the available evidence. It is possible that all the emeralds are *green*, and it is also possible that all emeralds are *grue* and that you've seen green coloured ones but not the blue coloured ones yet. But these properties also make opposite predictions about what colour the next emerald you pick up will be. If *green* is true, then the next emerald will be green. If *grue* is true, then the next emerald will be blue. And since both are true, a clear prediction cannot really be made. And yet, of course we all predict that the next emerald will be green. Why? This is the problem of induction. Viewed in this way, induction is a problem because the available evidence can support many different and contradictory conclusions.

With the earlier problem of induction defined by Hume, the solution was straightforward. Hume stated that we have a habit to make inductions. And our current understanding of learning theory suggests that we naturally generalise. Goodman's problem of induction is more subtle because it assumes that we do have this habit. If we have a habit to make inductions, how do we choose which one of the two possible inductions to make in the emerald example? A possible solution is that some ideas, descriptors and concepts are entrenched in our language and concepts and thus more likely to be the source of our inductions. Entrenchment means that a term or a property has a history of usage within a culture or language. And as we discussed earlier in Chapter 10, there is considerable evidence that language can influence and direct thoughts. In the emerald example, *green* is an entrenched term. *Green* is a term that we can use to describe many things. It is a basic colour term in English. It has a history of usage within our language of being used to describe many different categories of things. *Green* is therefore a useful property to make predictions from and about. By saying that a collection of things (emeralds) is green, we can describe all the things. *Grue*, on the other hand, is not entrenched. There is no history of usage and no general property of *grue* outside the emeralds that were *grue* yesterday and blue tomorrow. Unlike *green*, *grue* is not a basic colour term and does not apply to whole categories. Goodman argued that we can only make reliable inductions from entrenched terms and from coherent categories, and from natural kinds.

The term 'natural kinds' comes from philosophy also, specifically the work of Willard Van Orman Quine (Quine, 1969). Quine argued that these natural kinds are natural groupings of entities that possess similar properties, much like what we have referred to earlier as a family resemblance concept. Quine argues that things form a 'kind' only if they have properties that can be projected to other members. For example, an *apple* is a natural kind. This is a natural grouping, and what we know about apples can be projected to other apples. 'Not apple' is not a natural kind because the category is simply too broad to be projectable. This grouping consists of everything in the universe that is not an apple. Quine argued that all humans make use of natural kinds. Our concepts are formed around natural kinds. Our ideas reflect natural kinds. And reliable inductions come from natural kinds.

Granny Smith apples and Gala apples are pretty similar to each other and belong to the same natural kind concept. Much of what you know about Granny Smith apples can be projected to Gala apples with some confidence, and vice versa. The same would not be true of Gala apples and a red ball. True, they may be similar to each other on the surface, but they do not form a natural kind. Whatever you learn about the Gala apple, can't be reliably projected to the red ball. Note that this also bears on the example from the beginning of the chapter, in which I compared a Hubbard squash to a rugby football. They may be the same size and shape, but it ends there. Hubbard and pumpkins are a natural kind. We know this and so we use it to generate good, reliable, trustworthy inferences. But Hubbard and rugby football are not a natural kind. We can note the surface level similarity, but not project any other properties.

Quine's idea of a natural kind suggests the solution for Nelson Goodman's problem of induction. In fact, that was the point of his essay. Quine pointed out that *green* is a natural property and that green emeralds are a natural kind. Because of this, the property of green can be projected to all possible emeralds. *Grue*, being arbitrary and unstable, is not a natural kind and cannot be extended to all possible members. In other words, green emeralds form a kind via similarity; *grue* emeralds do not. Green emeralds are a coherent category, whereas grue emeralds are not. Green and grue might both be technically true, but only one of them is a coherent category, and natural kind, and a group with a consistent perceptual feature. As a result, we are able to make inductions about green emeralds and do not consider making inductions about grue emeralds.

Categorical Induction

Taking the research and the philosophy above, we can conclude the following. First, most (maybe all) organisms have a tendency to display stimulus generalisation. This can be as simple as basic conditioning or generalising about a group of people. Second, basic stimulus generalisation is universal and is sensitive to the similarity between the current stimulus and mental representations of previously experienced stimuli. Third, as we know from the research discussed in Chapters 8 and 9, concepts and categories are often held together by similarity. As a result, a productive way to investigate inductive reasoning is to consider that inductions are often based on concepts and categories. This is known in the literature as categorical induction. By assuming that induction is categorical, we make an assumption that there is a systematic way in which the past influences future behaviour. The past influences the present and judgements about the future as functions of our conceptual structure. That's the extraordinary predictive power of concepts!

Let's define categorical induction as the process by which people arrive at a conclusion or at the confidence about whether a conclusion category has some feature or predicate after being told that one or more premise categories possess that feature or predicate. This is just like our ongoing example about winter squashes. If you learn that a winter squash has fibres and large seeds inside, and then learn that the Hubbard squash is also a winter squash, you use your knowledge about the category of winter squash and the features that are typical of that category to make the inductive inference. In this way, the conceptual structure of past knowledge influences your prediction that the Hubbard squash will also have seeds.

In many of the examples I discuss below, the induction is made in the form of an argument. Not an argument like people arguing with each other, or an argument between ideas. But an argument as a statement with one or more premise statements that support a conclusion (i.e., the inductive inference). A premise is a statement of fact about something, someone, or a whole class. The premise contains predicates, which can be things and properties. In most of the examples, the predicates are properties or features that are common to the category members. The inductive argument also contains a conclusion statement. The conclusion is the actual inductive inference, and it usually concerns possible projection of a predicate to some conclusion object or category. In an inductive argument, participants would be asked to decide whether or not they agreed with the conclusion. They might be asked to consider two arguments and decide which of the two is stronger. For example, consider the inductive argument below, which first appeared in Sloman and Lagnado (Sloman & Lagnado, 2005).

Premise: Boys use GABA as a neurotransmitter.
Conclusion: Therefore, girls use GABA as a neurotransmitter.

The first statement about boys using GABA as a neurotransmitter is a premise. Boys are a category and the phrase 'use GABA as a neurotransmitter' is a predicate or fact about boys. How strongly do you feel about this conclusion? Part of how you assess the strength has to do with whether or not you think girls are sufficiently similar to boys. In this instance, you probably agree that they are pretty similar with respect to neurobiology, and therefore you would endorse the conclusion.

When you were answering this question, you may wonder what GABA is beyond being a neurotransmitter. You may not have actually known what it is at all and may not have known whether or not it is present in boys and girls. The answer to this conclusion is initially unknown. The statement is designed this way for a reason. The categorical induction statement works because it asks you to infer a property based on category similarity, rather than by retrieving the property from semantic memory. Thus, in the example above, GABA is a blank predicate. It is a predicate because it is the property we wish to project. But it is blank because we do not assume to know the answer. It is plausible, but not immediately known. And because you cannot rely on your factual knowledge about GABA as a neurotransmitter, you have to make an inductive inference on the basis of your knowledge about the categories (boys and girls in this case). This arrangement forces the participant to rely solely on categorical knowledge and induction, rather than on the retrieval of a fact from semantic memory.

Using this basic paradigm as an example, we can explore some general phenomena about categorical induction. Let's look at a few of these. Keep in mind that these phenomena tell us about how induction works and by extension, how concepts, categories, and similarity influence thinking and behaviour.

Similarity effects

For example, if the facts and features in the premise and the conclusion are similar to each other, are from similar categories, or are from the same category, inductive inferences can be made confidently. This is referred to as premise-conclusion similarity. According to Daniel Osherson and colleagues, who defined a theory of induction known as the 'similarity-coverage theory', arguments are strong to the extent that the categories in the premises are similar to the categories in the conclusion (Osherson, Smith, Wilkie, López & Shafir, 1990). We are more likely to make inductive inferences between similar premise and conclusion categories. This is fundamental, and aligns with Shepard, Hume, Goodman and all the rest. For example, consider the following two arguments:

Argument 1

Premise: Robins have a high concentration of potassium in their bones.

Conclusion: Sparrows have a high concentration of potassium in their bones.

Argument 2

Premise: Ostriches have a high concentration of potassium in their bones.

Conclusion: Sparrows have a high concentration of potassium in their bones.

In this example, the *high concentration of potassium in their bones* is a blank predicate. Which seems like the better argument? Which of these seems like the stronger conclusion? There's no 'right' answer, but Argument 1 should seem stronger and in Osherson's empirical studies, research participants find this to be a stronger argument. That's because robins and sparrows are similar to each other; ostriches and sparrows are not similar. The low similarity between the ostrich and the sparrow is evident on the surface, as is the high similarity between the robin and the sparrow. We assume that if the robin and the sparrow share observable features, they may also share non-observable features like the concentration of potassium in the bones. We're more likely to notice that robins and sparrows are part of the same natural kind, even though we know ostrich and sparrow are in the same category.[39]

Typicality

The example above emphasised the role of similarity between the premise and the conclusion, but in the strong similarity case you may have also noticed that the robin is a very typical category member. The robin is one of the most typical of all birds. And remember that typical exemplars share many features with other category members. Typical category members have a strong family resemblance with other category members. And they can also be said to cover a wide area of the category space. What is true about robins is true of many exemplars in the bird category.

Premise typicality can affect inductions about the whole category. For example, consider the following set of arguments:

Argument 1
> **Premise:** Robins have a high concentration of potassium in their bones.
> **Conclusion:** All birds have a high concentration of potassium in their bones.

Argument 2
> **Premise:** Penguins have a high concentration of potassium in their bones.
> **Conclusion:** All birds have a high concentration of potassium in their bones.

In this case, you might agree that the first argument seems stronger. It is easier to draw a conclusion about *all birds* when you are reasoning from a typical bird like a *robin*, which covers much of the bird category, than from a very

---

39 This should seem like a repudiation of the classical view of concepts described in Chapter 9. Robins, sparrows and ostriches are all part of the same category, but the featural overlap and stronger family resemblance of the robin-sparrow argument will cause people to find that to be the better argument.

atypical bird like *penguin*, which does not cover very much of this category. If we know that a penguin is not very typical – it possesses many unique features and does not cover very much of the bird category – we are not likely to project additional penguin features onto the rest of the category. We know that many penguin features do not transfer to the rest of the category.

Diversity

The preceding example suggests a strong role for typicality, because typical exemplars cover a broad range of category exemplars. But there are other things that can affect the coverage as well. For example, the diversity effect comes about when several premises are dissimilar to each other. Not completely unrelated, of course, but dissimilar and still in the same category. When presented with two dissimilar premises from the same category, it can enhance the coverage within that category. For example, consider the two arguments below (again from Osherson):

Argument 1
    Premise: Lions and hamsters have a high concentration of potassium in their bones.
    Conclusion: Therefore, all mammals have a high concentration of potassium in their bones.

Argument 2
    Premise: Lions and tigers have a high concentration of potassium in their bones.
    Conclusion: Therefore, all mammals have a high concentration of potassium in their bones.

Looking at both of these arguments, it should seem that statement one is a stronger argument. Indeed, subjects tend to choose arguments like this as being stronger. The reason is that lions and hamsters are very different from each other, but they are still members of the same superordinate category of mammals. If something as different and distinct as the lion and a hamster have something in common, then we are likely to infer that all members of the superordinate category of mammal have the same property. On the other hand, lions and tigers are quite similar in that both are big cats, both appear in the zoo in similar environments, and they co-occur in speech and printed text very often. In short, they are not very different from each other. And because of that we are less likely to project the property of potassium in the bones to all mammals and are more likely to think that this is a property of big cats, or cats in general, but not all mammals. The diversity affect comes about because the diverse premises cover a significant portion of the superordinate category.

The inclusion fallacy

Sometimes, the tendency to rely on similarity when making inductions even produces fallacious conclusions. One example is known as the inclusion fallacy (Shafir, Smith & Osherson, 1990). In general, we tend to prefer conclusions in which there is a strong similarity relation between the premise and the conclusion category. We tend to discount conclusions for which there is not a strong similarity relation between the premise and the conclusion. Usually, this tendency leads to correct inductions, but occasionally it can lead to false inductions. Take a look at the statements below and think about which one seems like a stronger argument.

Argument 1
   Premise: Robins have sesamoid bones.
   Conclusion: Therefore, all birds have sesamoid bones.

Argument 2
   Premise: Robins have sesamoid bones.
   Conclusion: Therefore, ostriches have sesamoid bones.

Which argument seems stronger? It's probably clear enough that the first argument seems stronger. Robins are very typical members of the bird category and we know they share many properties with other members of the bird category. It seems reasonable to infer that if robins possess sesamoid bones, so do all other birds. Most people find the second statement to be less compelling. Robins are typical, but ostriches are not. We know that robins and ostriches differ in many ways, so we are less willing to project the property of sesamoid bones from robins to ostriches. So, this probably strikes us as reasonable and fair. And not as a fallacy. But it is a fallacy and another example of how our intuitions seem to guide us to made inferences that *seem* obvious but may not be correct.

The reason why this is a fallacy is that all ostriches are included in the 'all birds' statement. In other words, if we are willing to accept Argument 1 that a property present in robins is also present in all birds, then that inference includes ostriches already. It cannot be the case that a single member of the 'all birds' category is a less compelling argument than all birds. If we are willing to project the property to an entire category, it is not correct to assume that specific members of that entire category do not have that property. Otherwise, we should not be willing to accept the first argument.

But most people find the first argument to be more compelling because of the strong similarity of robins to other birds. Robins possesses features that are common to many other birds. We recognise that similarity and judge the inference to all birds accordingly. The atypicality of the ostrich undermines the argument. People are likely to use similarity relations rather than category inclusion when making these kinds of arguments. Similarity seems to be the stronger predictor of inferences. Category membership is important, but featural overlap may be even more important. As with previous examples, this also undermines the classical view of concepts and supports a probabilistic, family resemblance view instead.

Category coherence

Inductions can be made from concepts and categories on the basis of similarity between premises and the conclusion, but the character of the concept also plays a role. One of the ways to see this is to consider the role of category coherence. The coherence of a category is related to how well the entities in the category seem to go together. For example, *firefighter* seems to be a coherent category. We expect there to be a high degree of similarity among people who join the fire department, and we might expect to them to share features, traits and behaviours. Upon hearing that someone is a firefighter, we might feel confident in our predictions about how they might act and behave.

But not all categories are that coherent. For example, *restaurant waiter* might seem much less coherent. Compared to firefighter, there is probably more diversity in this category and more possible reasons why people would have the job. Maybe there is more variability in appearance, behaviour and we are less likely to predict how people might act or behave as a function of their being waiters. In other words, there may be a coherence effect in categorical induction in which we prefer to reason from the most coherent category available.

This was studied directly by Andrea Patalano and her colleagues (Patalano, Chin-Parker & Ross, 2006). In their research, they considered social/occupational categories that were higher or lower in coherence. In order to make this determination, they first asked a group of research participants to rate categories on a construct called entitativity which is a measure that takes into account how members of a category are expected to be alike, how informative it is to know that a something is a member of a category, and whether or not members of a category might possess an inherent essence. Categories that are high in entitativity are thought to be highly coherent.

Patalano et al. found that categories like *soldier, feminist supporter* and *minister* were all highly coherent whereas *matchbook collector, country clerk* and *limousine driver* were low in coherence. Then the researchers carried out an induction task in which people were asked to make predictions about people who were members of more than one category. For example, imagine that the following information is true:

> **Premise:** 80 per cent of feminist supporters prefer Coca-Cola to Pepsi.
> **Premise:** 80 per cent of waiters prefer Pepsi to Coca-Cola.
> **Premise:** Chris is a feminist supporter and a waiter.
> **Conclusions:** What beverage does Chris prefer, Coca-Cola or Pepsi?

People were asked to make an inductive decision and to rate their confidence. They found that people in their experiments preferred to make inductions about the more coherent categories. That is, in the example above, people say that Chris was likely to prefer Coca-Cola. Because people view *feminist supporter* as the more coherent category, they preferred to base their inductions on that category.

Induction is a crucial and critical cognitive action. Without it, we would be lost among a sea of novel objects and novel properties, unable to use the past, unable to rely on memory. I argued at the outset that although these inductions can be the output of both System 1 or System 2, they are often automatic and unavoidable. They are often driven by System 1 associations and spreading activation. It's impossible *not* to use what you know to predict unseen properties, unknown features. Eliminating uncertainty is important for our survival. Any species that can gain even the smallest advantage by being able to predict, with more certainty than chance, what might come next is going to rely on that ability and prosper as a result. This is surely why we (as in all sentient and semi sentient beings) carry on this way, inducing and predicting. Induction is necessary for our survival.

There are some caveats to our inductive instinct. These caveats are related. First, we sometimes make mistakes. Sometimes we infer something that will not turn out to be true, or a property that does not actually exist. Induction is probabilistic by nature. Most apples are sweet, but sometimes they are sour. By nature, our inductions are not certain. And yet we treat them with certainty. It's our risk, our gamble. We benefit by being sometimes, occasionally wrong because it lets us be faster. There is a second caveat. Because the inductions are sometimes, occasionally wrong, we need to be sure not to let them interfere with our interactions, plans and behaviours. We may very well be predisposed by nature to generalisation, induction, stereotyping and prejudice. But we need not be predisposed to bigotry, racism and hated. Depending on our circumstance, we may need to override these impulses, recognise and correct a bias, and lean more heavily on System 2 to override System 1. For that too, is necessary for our survival.

One of many ways to be more certain is to rely on more carefully constructed arguments when needed. We do have a system for that. It's called deduction. Deduction, properly carried out, allows us to think in ways that are able to arrive at true and valid conclusion. Induction lets us quickly and probabilistically predict the future. Deduction lets us ferret out what's true.

And you can predict, based on how this paragraph is ending, that I'm going to talk about deduction next.

CHAPTER 13

Deducing the Truth

W hen my kids were young, sometimes they would lose something important, or misplace something, like a coat, a book or a phone. They would get home from school and say, 'I can't find my jacket'. I'd grumble. Then I'd say, 'where was the last place you saw it?', and then we'd talk about retracing steps and trying to remember where the thing was last seen. I might say, 'Well, if it's not in your backpack, then it must be at school'. This is standard parent talk, but by saying that, I'm essentially suggesting that we frame this as deductive logic. I start with the premise that it must be at one location or another and then look to verify the conclusions. It can only be in one place, and we can rule out options and narrow the search. If we look everywhere at home, and do not find the missing item, we conclude that it must be at school. We can make that conclusion even if we don't actually see it. We can verify the conclusion the next day. My kids might not realise it, but they are working through a basic deduction problem. Sometimes this deductive process results in finding the missing item.

Deduction vs Induction

Inductive reasoning involves making a predictive inference from observations. We can call this a 'bottom up' style of reasoning because induction moves from the specific (observations) to general (conclusions based on evidence). When we rely on induction, we make conclusions that are likely to be true, that are probably true, or that we have some confidence in. In other words, the outcomes are probabilistic. Both induction and deduction involve going beyond the given evidence to discover something new via thinking. With deduction, however, we are often making a specific conclusion and trying to determine if that conclusion is valid. Deduction usually begins with a general statement ('My jacket is either in my backpack or at school'), and then proceeds to more specific statements ('It is not in my backpack') that can add new information, support or not support the initial statement, and eventually lead to a conclusion ('my jacket must be at school').

The two kinds of reasoning are related. We rely on both induction and deduction in our everyday thinking. We use them together so often that it can be difficult to tell if you are using inductive reasoning or deductive reasoning. For example, suppose I buy a coffee from a Starbucks coffee store, take a sip, and discover that it is very hot. I conclude that *Starbucks coffee is very hot*. That is an observation and a generalisation. Using induction, I can infer that other Starbucks stores will serve very hot coffee. Categorical induction might even allow me to project and extend the property of 'hot' to coffee purchased at other, similar restaurants and stores. In both cases, I am relying on my observations, experience, and inductive reasoning to make predictions about the future.

So how does deductive logic differ from that process? It has to do with the nature of the conclusions and the nature of the relationships between the premises and the conclusion. Let's look at these as formal statements. Suppose I form a general belief about Starbucks coffee from all my experiences. Let's call that general belief a *premise* which can be stated formally as:

> **Premise**: All Starbucks coffee is very hot.

This premise is a statement of fact about something I know, was taught or can verify. This is not a conclusion, though it may have been derived from an inductive inference at some point. The premise does not have the same kind of nuance that an inductive inference has. It's just a statement. In this case, it's a fact about Starbuck's coffee. A premise like this can be combined with other premises and can make precise conclusions. The combination of premises and a conclusion together is called a 'syllogism'. For example:

> **Premise**: All Starbucks coffee is very hot.
> **Premise**: This coffee is from Starbucks.
> **Conclusion**: Therefore, this coffee is very hot.

In a syllogism like this, we assume that the premises are true or that we believe them to be true. In the example above, the conclusion is valid, because it's the only possible conclusion that follows from the premises. That's the definition of a valid deduction. A *valid conclusion* is one for which the conclusion necessarily follows from the premises and for which there are no other possible conclusions. If these true premises allowed for an alternative conclusion, then the deduction would not be a valid one. When the premises lead to a single conclusion like this, it's like a checkmate in chess, or like a putting the final piece in a puzzle. Everything fits. Everything works, and you can feel confident in the conclusion. Thanks to deduction, you just discovered something new by thinking about it!

But what if one or more of the premises is not true? That's like building a house on shaky foundations. The logic might lead to a valid conclusion, but it may end up conflicting with other facts in the world. In addition to validity, we also care about the soundness of a syllogism. We need to consider the soundness of a deductive argument. A *sound argument* is one that is valid (because only one conclusion can be drawn from the premises) and one for which the premises are also known to be true. These two ideas, validity and soundness, are important in making deductive conclusions. It is possible to have a valid deduction that is not sound. In the case above, the argument is only a sound one if we know that 'All Starbucks coffee is hot' is true. If there is evidence to the contrary, the argument can still be valid in its construction, but it is not a sound deduction, and so we may not trust the conclusions.

So, now that you know what deductive logic is, how it is different from inductive reasoning, and what counts as valid and sound, let's look more closely at the structure of logic. Logical structure is important in deductive reasoning because it's the structure of a task that determines its validity. Then I'll discuss some examples of how deduction is used in some common contexts and then cover some more complex examples. And I'll need to discuss how and why people often fail to reason logically. Logic is important, powerful, and possible. But we often fail to follow through with the work of verifying the soundness and the validity of the argument and rely instead on our general knowledge and familiarity around the situation. In other words, we fall back on our heuristics, our biases, and on the faster 'system 1' thinking that characterises so much of the way we think.

The Structure of a Logical Task

As we've seen, you can use deductive logic to arrive at a conclusion about a member of a category (the hot coffee from Starbucks). If it's a sound argument, you're going to arrive at a conclusion that you are confident in. We can also rely on deductive reasoning to make predictions about all kinds of options and outcomes. For example, imagine that you are planning to go to a shopping centre with your friend. They text you to say that they will meet you by the Starbucks or by the ice cream shop. Either one or the other. You have two options, and they cannot both be true at the same time. For that matter, they cannot both be false either. According to your friend, they will meet you at one place or the other. One will be true, and one will be false. Or rather, your friend will be at one place and not the other. How can we use logic to predict where they will be? First, let's set up a syllogism:

> **Premise:** Your friend is waiting at the Starbucks or at the ice cream shop.
> **Premise:** Your friend is not at the Starbucks.
> **Conclusion:** Therefore, your friend is at the ice cream shop.

This deductive argument has several components. As with the previous examples, the statement has some premises and a conclusion. The premises give basic facts that can be confirmed (or not) by other premises. Unless we hear otherwise, we assume that the premises are true. That is a crucial aspect of deductive logic, because, in many ways, the challenge of deductive logic is evaluating the validity structure of the task.

So how do you form a premise? A premise is made of facts and operators. Facts are statements about things that can be true or false. Facts are descriptions or statements about an object's properties. The operators tell you about the relationship of the facts to other facts. These operators are crucial to the deduction task, and this is what makes deductive reasoning distinct from inductive reasoning. In the example above, the operator 'OR' adds a layer of meaning to the statement. This is what allows us to think *conditionally* about the two alternatives. Your friend is either here *OR* there. The operators let us operate on the information that we have. They let us think formally. In the premise above, one must be true, but both of these cannot be simultaneously true. If you are given information that one of these is false (not at the Starbucks), then you can conclude that the other must be true (at the ice cream shop). In the second premise, 'NOT' is also an operator that tells us that one half of the first premise is not true.

And of course, deductive problems also have conclusions. That's the main reason we have deductions. These conclusions can be delineated with expressions like 'therefore' or 'then'. Assuming the premises are true, the conclusion is either valid or not valid. A conclusion is valid if it is the only possible conclusion given the truth of the premises. A conclusion is not valid in cases where the same set of premises can give rise to more than one conclusion. A valid deductive argument is a really strong thing. This way of thinking lets us make conclusions with certainty and confidence.

Deductive reasoning and deductive logic are powerful. And we can think logically, given the right circumstances. But we often fail to reason logically. Instead, we tend to use our memories and heuristics. I have made this point throughout this book. We use heuristics because they are faster and usually correct. We rely on System 1 thinking because it's easier. It's effective. It's adaptive. But like many biases, if we understand a bit more about how and why they work, we can learn to recognise them and avoid them when they might lead us into making the wrong inference, decision or conclusion.

We are capable of using logic and of reasoning. We are capable of deducing the truth. We don't always do it properly, though, because it can be time consuming and resource intensive. So, let's take a deeper look at how deduction works and how we reason correctly. Then we'll take a look at some of these reasoning biases and how to avoid them.

Categorical Reasoning

I started the chapter with an example about how we might notice that Starbucks coffee is very hot. We can use that information to deduce something about a specific cup of Starbucks coffee. Although the conclusion in that example is about a single cup of coffee, the argument works because we're making assumptions about *all* the coffee from Starbucks. We started with a premise that 'all Starbucks coffee is very hot'. This is important because when we make a statement like that, we're assuming that every member of the category has that property. This is categorical reasoning and it is built upon the basics of categorisation and knowledge that I discussed earlier.

Categorical reasoning is a form of logical deduction that occurs when we make conclusions that depend upon category membership. Categorical reasoning is sometimes also referred to as 'classical reasoning', because we are reasoning about a class of things. The terms are interchangeable. In fact, we can even express this as a premise: 'all classical reasoning is categorical'. If something belongs to a class or a category, then we can probably make a categorical statement about it. And by the way, the statement I just made is an 'if/then' statement and that's another kind of logical statement that I'm going to discuss in a few pages. It's possible to make all sorts of premise statements about things: All coffee is hot. All coffee is liquid. This coffee is hot. Some coffee is bad. Premises are fairly straightforward to state. The more challenging part of categorical reasoning is figuring out how to combine different premises to arrive at a valid conclusion.

In the previous chapter on inductive reasoning, I also emphasised the importance of categories. In that case, the emphasis was on the similarity between the premise and the conclusion. Stronger similarity among premises often results in stronger inductions. With deduction, the emphasis is on the actual category membership rather than similarity. Let's look at another example, in the same form as my 'hot Starbucks coffee' example.

First premise: All men are mortal.
Second premise: Socrates is a man.
Conclusion: Therefore, Socrates is mortal.

In this classical syllogism, we have a first premise which is a general statement. This is a statement about the category. In this case, we are suggesting that the category (men) is included in or equivalent to another category (mortal things). There is an overlap between the category of men and the category of mortal. The first premise tells us about the relationship between the two so that we can transfer the properties of one category to the other. In the second premise, the statement provides specific information about members of the categories. Since all men are mortal and we know that Socrates is a member of the category of men, we can conclude that he is also a member of the mortal category, because of the relationship between the two categories stated in the first premise. Notice that there is little role for similarity or featural overlap in these statements. It does not matter how similar (or not) Socrates is to the category of men. All that matters for the argument is that we know that he is a man.

This is simplistic example. It makes a universal statement about the relationship between two categories. But the nature of the relationship between the two categories is never made clear. We don't know if 'men' is a subset of 'mortal things' or if the two categories are entirely the same. Does this matter? In some cases, it does matter. In order to make more sense out of this, let's look at four ways to express the relationship among categories.

The universal affirmative is a statement about the positive relationship between two categories that is universal for all members. I say that 'all cats are mammals', I am making a universal affirmative form. It's universal for cats (it concerns *all* cats). It's affirmative for cats as mammals (they *are* mammals). We've seen this example a few times: 'All Starbucks coffee is hot', 'All men are mortal'. It's easier to discuss the form of the statements by replacing the concepts with variables: 'All As are Bs'. This is a universal affirmative form that describes all of the examples. Everything in category A is also included in category B.

One interesting aspect of this universal affirmative form is that it is not reflexive. It has two possible interpretations. In one interpretation, all the As are Bs and all the Bs are also As. There are some examples of this, but they are not easy to come up with and they usually involve synonyms: 'all people are human' or 'all cars are automobiles' or 'all cats are felines' but these are not really very informative and it's hard to think of examples beyond these synonymous ones. Can you think of any good examples? I can't.

The other interpretation for this is the case where all the members of category A are also members of category B but as a subset of category B. In this version, category B might be a superordinate category. This would be like saying 'all cats are animals' or 'all cars are vehicles'. This tells you a great deal about the first concept and some information about the second. It's still universal and affirmative for the first concept. It tells us that everything in category A is also part of category B and therefore will inherit some properties of category B. But a premise like this does not tell us about all of the members of the second concept. There might be other members of category B that do not overlap with category A. In other words, all cats are animals, we can agree. But not all animals are cats.

Before I go on to discuss more complex premises, let's pause to look at a diagram. Whenever I think about logical deductions, I like to imagine the concepts and categories as circle diagrams. In the figure below, I have drawn out the two different possible arrangements for the universal affirmative (*Figure 13.1*). I usually call these 'circle diagrams' but they are more properly known as 'Euler diagrams'. You might also call them 'Venn diagrams' but that's not correct. The Venn diagram also uses colour or shading to illustrate the strength of the overlap. Let's just stick with 'circle diagrams' because they are simpler to work worth.

On the left, I've drawn out the case where all the As belong to B as a subset, or 'All A are also B'. Every A is a B, but not every B is an A. Imagine that the circles encompass all the possible cases in the universe that you know of. This suggests that all members of category A are contained within a larger category B. This suggests a hierarchical relationship such that B is the larger category and A is a subcategory. In this case, it is true to say that all A are B, but the reverse is not true. It is not true to say that all B are A. On the right is the other form of 'All A are B'. In this case, there is a reflexive relationship and everything in category A is the same as members in category B.

Universal Affirmative

All A are also B

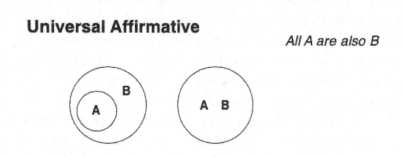

Particular Affirmative

Some A are also B

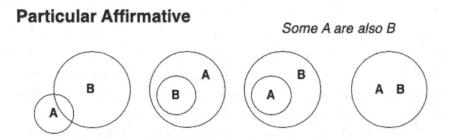

Figure 13.1: Two possible circle diagrams that illustrate the universal affirmative and four possible circle diagram to illustrate the particular affirmative.

Particular affirmative

If I say, 'some cats are friendly', I am using an expression known as a particular affirmative. We mean that some members of the cat category are also members of the category of friendly things. The particular affirmative suggests that some members of one category can also be members of another category. As shown in Figure 13.1, there are four possible versions of this statement. Considering the abstract version 'Some A are also B', the diagram on the left shows a case in which A and B partially overlap, which allows for 'Some A are also B' to be true in the intersection. The second of these four is a case where there is a superordinate A and a subordinate B, which allows for 'Some A are also B' because all the B are A, but not all the A are B. The statement is still true in this universe.

The next two diagrams are more difficult conceptually. When we hear that 'Some A are also B', it is important to realise that the word 'some' can mean *at least one, and possibly all*. The reason is that as long as one member of category A is also a member of category B, the statement is true. Even if all the As are also members of category B, the statement 'Some A are also B', is still true. Or think about it this way, if it turned out that all cats were friendly, and I said to you 'some cats are friendly', I would not be telling a lie. This would be a true statement. 'Some' does not preclude 'all'.

The third and fourth diagrams show two ways in which all of the As are also members of B. In both of these cases, the statements 'Some A are also B' is still true. Although this is an example of a universal affirmative for A, it can also be an example of a particular affirmative. Admittedly, it is an incomplete statement but in a universe in which all members of category A are equivalent to all members of category B, it does not mean that the statement 'Some A are also B' is false.

What this means is that the particular affirmative is a more difficult statement to evaluate. It gives reliable information about the status of at least one and possibly all members of category A. It tells you very little about the status of category B and it tells you very little about the entire relationship between categories A and B. When evaluating a series of statements for which one of them is a particular affirmative, considerable care must be taken to avoid an invalid conclusion.

Universal negative

If we say that 'no cats are dogs', we are using a statement referred to as a universal negative. The universal negative expresses a relationship between two concepts (A and B) for which there is absolutely no overlap. In contrast to the universal affirmative and the particular affirmative, the universal negative has only one representation. It is also reflexive. The statement does not tell us much about other relationships with respect to category A and category B, except to tell us that these two categories do not overlap at all (see *Figure 13.2*).

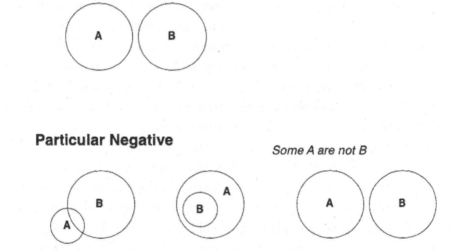

Figure 13.2: One possible circle diagram that illustrates the universal negative and three possible circle diagrams to illustrate the particular negative.

Particular negative

If I say that 'some cats are not friendly', I'm using the particular negative. In this case, I want to get across the point that some members of one category are not members of another category. As with many of the other examples, there are several ways for this statement to be true. In Figure 13.2, we see three different ways in which the statement 'Some A are not B' can be true. The case in the bottom left shows A and B partially overlapping. This allows for the statement to be true, because there are members of category A that are not contained within category B. In the bottom centre, B is a subordinate group contained within A. This also allows for the statement to be true, because there are many of category A that are not contained within the subordinate category B. In the example at the bottom right of the figure, a diagram shows a case where the two categories do not overlap at all. Although this diagram also represents a universal negative relationship, the statement 'Some A are not B' is still technically true in this case. If it were the case that there were no friendly cats anywhere (easily disproven by my cat, Peppermint) and I said 'some cats are not friendly' my statement would be still true in this no-friendly-cat universe.

Errors in categorical reasoning

Reasoning about categories and concepts is a fairly common behaviour, but because of the occasional ambiguity and complexity of these classical relationships, people often make errors. In addition, many of the errors that we make are a result of conflating personal beliefs and knowledge with the notion of logical validity. One way to avoid making these errors is to use the simple circle diagrams like those shown in Figures 13.1 and 13.2 to determine whether or not a conclusion is valid. If there is more than one configuration that allows the premises to be true but that lead to different conclusions, then it is not a valid deduction. Consider the following syllogism:

Premise: All doctors are professional people.
Premise: Some professional people are rich.
Conclusion: Therefore, some doctors are rich.

The first premise tells us something about the relationship between doctors and the category of professional people. It tells us that everyone who is a doctor is also a member of this category. It leaves open the possibility that the two categories are entirely overlapping, or that there is a larger category of professional people which also includes teachers, engineers, lawyers, etc. The second premise tells us something about some of the professional people. It tells us that some of them (meaning at least one and possibly all) are rich. Both premises express a fact and in both cases the fact conforms to our beliefs. We know that doctors are professional, and we also know that at least some people in the category of professional people can be rich.

The conclusion that we are asked to accept is that some doctors are also rich. The problem with this deduction is that it conforms to our beliefs and those beliefs can interfere with our ability to reason logically. And if it conforms to beliefs that we are familiar with, we're liable to rely on System 1 and our knowledge instead of System 2 and logic. We might know a rich doctor or have friends or family who are rich doctors. That is not an unreasonable thing to believe because although not all doctors are rich, certainly we are all aware that some of them can be. We know this to be true from personal experience, but that knowledge does not guarantee a valid deduction. A conclusion is valid only if it is the only one that we can draw from the stated premises. This is an example of the belief bias I discussed earlier, in which people find it easier to evaluate believable statements as being valid and non-believable statements as being invalid. In this case, the conclusion is not valid, but it's believable.

Figure 13.3 shows two of several possible arrangements of the categories. Each of these arrangements allow the premises to be true, but each arrangement ends up arriving at a different conclusion. On the left, the diagram shows doctors as a subcategory of professional people, and it shows a category of rich people that overlaps with professional people and includes the doctors. In this universe, the first premise is true because all of the doctors are professional people. The category of rich people overlaps

Belief Bias

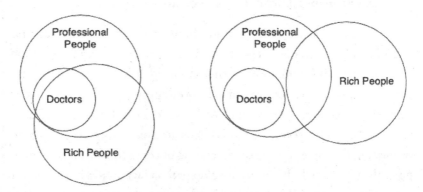

Figure 13.3: This is an example of an invalid categorical statement and the belief bias effect. Both sets of circles allow for the premises to be true and yet they support contradictory conclusions.

partially with the category of professional people, allowing the second premise to be true. Finally, because the category of rich people overlaps partially with the category of professional people and includes the doctors, it allows for the conclusion to be true as well. This state of affairs conforms to our understanding of doctors as generally being financially well-off.

The problem is that this is only one of many possible arrangements of the classes that allows for the premises to be true. On the right is an alternative arrangement. In this case, the category of doctors is still completely subsumed within the category of professional people, thus allowing for the first premise to be true. This also shows that the category of rich people partially overlaps with the category of professional people, thus allowing for the second premise to be true. However, the overlap between the rich people and the professional people excludes all of the doctors. In this arrangement, both premises are still true, but the conclusion, 'Therefore, some doctors are rich', is not. In this arrangement, no doctors are rich. The existence of both of these arrangements, each of which allow for the premises to be true but make different predictions with respect to the conclusion, indicates that this is not a valid syllogism.

Of course, you might object to the conclusion of invalidity because it is probably not true that no doctors are rich. You surely know one or two rich doctors or at least have heard of one or two rich doctors. In doing so, you would be showing a belief bias. That is one of the things that is challenging about deductive logic. This is not a valid argument and yet the conclusion can still be true. In logical deduction, it is often difficult to separate truth from validity.

We have a tendency, or a bias, to assume that conclusions are valid if they align with something we already believe and that they are not valid if they do not align with something we already believe. Although we are constantly making inferences, drawing conclusions, and making predictions about things, deductive logic can often seem counterintuitive if it does not line up with what we believe to be true. We often agree with a conclusion and think that it is valid, even if it is not. We may reject conclusions that are valid. This is a bias, because validity is determined by the structure of a logical task, and not its believability. But it's also understandable, because we have a tendency to rely on concepts and memory to make conclusions. That is, we have a tendency to make decisions based on System 1, the fast system I discussed earlier in Chapter 11. This fast system is helpful because it lets us make decisions and deductions quickly, but it also has a tendency to encourage biases like this.

Conditional Reasoning

In the previous section, I considered reasoning about classes of things, but people also reason about conditionality and causality. These kinds of deductions are usually framed within the context of if/then statements. For example, 'if you study for the exam, then you will do well'. This statement reflects a relationship between a behaviour (studying) and an outcome (doing well). It reflects only one direction; there may be other things that affect your doing well. And as with categorical reasoning, there are several forms of conditional reasoning. The combination of these forms allows for a variety of valid and not valid statements to be expressed and evaluated.

Before describing the different versions of conditional reasoning, consider the components of a conditional reasoning statement:

> **Premise:** If A, then B.
> **Premise:** A is true.
> **Conclusion:** Therefore, B is true.

In the first premise, 'A' is referred to as the antecedent. It is the thing or fact that occurs first. 'B' is referred to as the consequent. The consequent is the thing that happens as a consequence of A being true. Although this might seem causal, it does not necessarily have to be. That is, we don't need to assume that A causes B, only that if A is true, B is true also. The second premise gives information about the antecedent within the premise. In this example, it gives information about the antecedent being true.

Affirming the antecedent

A common conditional argument is one in which a relationship is expressed between an antecedent and a consequent and then you are given information that the antecedent is true. Consider the example below:

> **Premise:** If my cat is hungry, then she eats her food.
> **Premise:** My cat is hungry.
> **Conclusion:** Therefore, she eats her food.

In this case, we are informed that if the cat is hungry (the antecedent), then she eats her food (the consequent). The second premise indicates that she is hungry and thus affirms the antecedent. Thus, it can be concluded that she will eat her food. If you accept these premises, you know that if the cat is hungry then she eats. This is a straightforward relationship that is sometimes referred to as *modus ponens*, which in Latin means 'the mode of affirming that which is true'. This deduction is valid. It is also easy for most of us to understand because it is consistent with our bias to look for confirmatory evidence. It is also easy to evaluate because it expresses things in the direction of cause and effect. Although conditional reasoning does not need to be causal, we still tend to think in terms of causal relationships. This chapter will discuss confirmation bias later.

Denying the consequent

In the previous example, the statement affirmed the antecedent to allow for a valid deduction. However, imagine the same initial premise and now the second premise denies the consequent:

> **Premise:** If my cat is hungry, then she eats her food.
> **Premise:** She does not eat her food.
> **Conclusion:** Therefore, my cat is not hungry.

In this example, the consequent is that she will eat her food. If that consequent is denied by saying 'She does not eat her food', then you can deduce that the antecedent did not happen. The first premise tells you the relationship between the antecedent and the consequent. If the antecedent occurs, then the consequent *must* happen. If the consequent did not happen, then it is valid to deduce that the antecedent did not happen either. This relationship is more difficult for most people to grasp. It runs counter to the bias to look for confirmatory evidence, even though it is still valid. This form is also known by a Latin name, *modus tollens*, which means 'the mode of denying'.

Denying the antecedent

When you affirm the antecedent or deny a consequent, you are carrying out a logically valid form of conditional reasoning. Both of these actions allow for a unique conclusion to be drawn from the premises. However, other premises produce invalid conclusions. For example, the statements below show an example of denying the antecedent:

> **Premise:** If my cat is hungry, then she eats her food.
> **Premise:** My cat is not hungry.
> **Conclusion:** Therefore, she does not eat her food.

In this example, the first premise is the same as the previous cases, but the second premise denies the antecedent by telling us that the cat is not hungry. With this information, you may be tempted to assume that the consequent will not happen either. After all, you are told that if she is hungry, she eats. You then find out she is not hungry, so it is natural to assume that she will not eat as a result. You cannot conclude this, however. The reason is that the first premise only gives us information about what happens with a *true* antecedent and a consequent. It does not tell you information about when the antecedent is not true. In other words, it does not rule out the possibility that the cat could eat her food for other reasons. She can eat even if she is not hungry. The cat could eat all the time, 24/7 regardless of hunger and the first premise is still true. Non hungry eating does not falsify the first premise. And so, finding out that she is not hungry does not allow you to conclude that she will not eat. You can suspect this. You can possibly infer that it might happen. But you cannot arrive at that conclusion exclusively.

Affirming the consequent

The final example is one where you receive information that the consequent is true. Just like the preceding example, this one might seem intuitive but is not logically valid. An example is below:

> **Premise:** If my cat is hungry, then she eats her food.
> **Premise:** My cat eats her food.
> **Conclusion:** Therefore, my cat is hungry.

The first premise is the same as the preceding examples and expresses the relationship between the hungry cat and the cat eating her food. The second premise affirms the consequent, that is, you are told that she does in fact eat her food. You may be tempted to infer backwards that the cat must have been hungry. But just like the preceding example, the first premise tells you a directional relationship between the hungry cat and eating but does not tell you anything at all about other possible antecedents for the cat eating her food. As a result, knowing that she eats her food (affirming the consequent) does not allow for an exclusive conclusion that the cat's hunger (the antecedent) was true. This is also an invalid deduction.

Confirmation Bias

Of all the biases I've discussed, one of the most frustrating is confirmation bias. You have almost certainly come across this bias before. This bias shows up any time we discount evidence that we don't agree with. This bias shows up when we seek out evidence that agrees with something we already believe. This bias is pervasive and ubiquitous. Earlier, I discussed an invalid categorical syllogism concerning some doctors being rich. If you believed that doctors were rich, you might show a confirmation bias if you only searched for evidence of rich doctors. The confirmation bias would also show up if you tended to downplay or discount information that was inconsistent with your belief, or that would disconfirm your belief. In other words, even if you did meet a doctor who was not rich, you might downplay that evidence as an anomaly, or someone who just wasn't rich yet.

We often see evidence of confirmation bias in popular media. In the 1990s, dietary advice strongly suggested that the best way to eat healthily and reduce weight was to reduce the amount of fat in the food you ate. There was a heavy emphasis on low-fat foods. At the same time, there was a heavy emphasis on eating high carbohydrate foods. Plain pasta was good; butter and oil were bad. Although we know now that this advice was not very sound, it had a long-lasting effect on personal health. One of the possible reasons is that when people were avoiding fat, they were avoiding higher calorie and richer foods. This may have given the impression that it was the fat that was causing dietary problems when in fact it may have been a simple issue of overall consumption. People notice many positive factors when switching to a restricted diet of any kind, such as vegan, ketogenic or so-called 'paleo' diets. If you switch to a diet like this, and you notice some weight loss, you tend to attribute the weight loss to the specifics of the diet rather than the general tendency to be more selective. This is a confirmation bias. You believe that eating a high protein diet will result in weight loss, and you may miss the alternative explanation that a restricted diet of any kind can also result in weight loss. The low-fat diet craze of the 1990s was even more difficult because of the strong but incorrect belief of the equivalency between dietary fat and body fat. It is possible that the surface-level correspondence between these two kinds of fats encouraged people to see a confirmatory match where one was not present.

The confirmation bias has often been studied with a psychological test called a 'card selection task'. In these tasks, research participants are typically given one or more rules to evaluate. The rules refer to the relationship of symbols, letters, numbers or facts that are presented on two-sided cards. In order to determine whether or not the rule is valid, subjects indicate which cards should be investigated. In this respect, the card selection tasks might have some degree of ecological validity with respect to how deduction is used in everyday thinking. The card task tries to answer the following question: If you are given a series of facts, how do you go about verifying whether or not those facts are true?

The most well-known example of a card selection task is the one developed by Wason in the 1960s (Wason, 1960). In this task, subjects are shown four cards laid out on a table. Each card can have a number or a letter on each side. Subjects are then given a rule or premise to evaluate and are told to indicate the minimum number of cards to turn over in order to verify whether or not this rule is true. For example, given four cards showing [A] [7] [4] [D], a rule might be:

> **Premise**: If a card has a vowel on one side then it has an even number on the other side.

Looking at the [A] [7] [4] [D] cards, which ones would you turn over in order to evaluate this rule? Most people agree that the first card to turn over would be the one with the [A] on it. If you turn this card over and there is no even number on the other side, then the rule is false. This is straightforward because it is an example of affirming the antecedent. The antecedent in this case is 'If a card has a vowel on one side', so you can see if that is true with the A card. You turn it over to see if the rule is being followed. In the original studies, Wason also found that people almost always suggested turning over the [4] card in addition to the A card. By turning over the [4] card, subjects were usually looking to see if there was a vowel on the other side.

Looking for a vowel in this case is an example of a confirmation bias where you look for evidence to confirm the statement. This is also an example of affirming the consequent, which is known to be an invalid form of conditional reasoning. The rule does not specify the entire range of possibilities with respect to even number cards. The even number can occur on the other side of a vowel card, as suggested by the rule, but the rule does not exclude the possibility of an even number occurring on the back of other cards. In fact, the rule would be true even if even numbers occurred on the back of all cards. If every single card shown in this array had an even number on the other side, the rule would be true.

Wason argued that the correct solution to this problem is to turn over the [A], and also the [7] card. The [7] card looks to disconfirm the rule. This is an example of denying the consequent. If the [7] card has a vowel on the other side, then the rule is false.

How does this bias come about? One possibility is a limitation in attention and working memory capacity and a tendency to rely on System 1. Given the statements, it may simply be a less demanding task to pick two cards that conform most closely to the hypothesis that was stated. Choosing to turn over a card that tests for denying the consequent requires the consideration of a premise that is not explicitly stated. In order to arrive at this implicitly stated premise, the subject must have sufficient working memory resources to hold the stated premise in mind along with the unstated premise. This is not impossible, but it may not be straightforward. As a result, people tend to choose confirmatory evidence.

In some ways, the pervasiveness of confirmation bias may be related to the notion of entrenchment that I discussed in Chapter 12. It is culturally and linguistically entrenched to think in terms of describing something 'that is'. So, when a person confirms a hypothesis, they look for evidence that something is true. The resulting search space is smaller and is constrained, and there is a direct correspondence between the hypothesis and the evidence. When searching for disconfirmatory evidence, the search space is much larger because people will be searching for something that 'is not'. Goodman (Goodman, 1983) and others have argued that 'what something is not' is not a projectable predicate. When thinking about categories, it makes sense to think about what something is, but not so much to think about what it is not. An animal can be described as a member of the DOG category, but it is not very informative to describe the same animal as a member of the NOT A FORK category, or the NOT A BEVERAGE category. The list of categories for which the animal is not a member is essentially infinite. So, with respect to category membership and reasoning, it is understandable that humans display a confirmation bias. Confirmatory evidence is manageable. Disconfirmatory evidence is potentially unmanageable.

Framing the task in a different way produces different results and can eliminate this confirmation bias. The standard confirmation bias shown in the Wason card selection task does not always play out. Alternative versions of the card selection task can be arranged that are formally equivalent on the surface but ask the subject to adopt a different perspective, such as what is permitted to be on a card, rather than an if-then statement. In many cases, a permission schema is easier for subjects to consider. When thinking about permission, it is common to think about what you can do and what you cannot do. Permission is something you are permitted or allowed to do, but we often conceive permission to be freedom from restriction. Speed limits tell us how fast we are permitted to travel, but we tend to consider the ramifications of exceeding the speed limit. The green light at a junction permits you to drive, but the bigger deal is what happens when the red light comes on and you have to stop.

Wason's card selection task can be reconstructed as one requiring permission. This is referred to as a deontic selection task. In this example, the cards have ages and beverages on either side.

Imagine four cards that say [21] [beer] [coke] [17].

Just as in the standard version, subjects are given a rule to evaluate and asked to indicate the minimum number of cards they need to turn over in order to evaluate this rule. In this case, the rule might be:

> **Premise:** If a person is drinking alcohol, they must be
> over eighteen.

Different countries have different minimum drinking ages, so just replace 'eighteen' with the age where you live when you consider this example. People rarely fail this task. It is straightforward to realise that you need to check the age of the beer drinker, and you need to do that checking with the person who is drinking. Even if you have never been in a scenario where you need to ensure that an establishment or club is following the law, most people know what it means to have legal permission to consume alcohol. Very few subjects show a confirmation bias here.

The explanation is that this task appeals to the permission schema. The permission schema essentially limits the number of hypotheses that need to be considered. It is important to note that this task succeeds in eliciting logical behaviour not because it makes it more concrete or realistic, but rather because the permission schema reduces the number of options and makes it easier to consider what violates the rule.

Deductive reasoning is fairly straightforward to describe in many ways. Deductive tasks generally follow a strict logical form. There are clear cases for deductions that are valid and not valid. And there is a fairly straightforward definition for sound and unsound deductions. And yet, most people have difficulty with deductive reasoning. Deductive reasoning seems to be outside the ability of many people. And as discussed in this chapter, many people reason, make decisions, and solve problems in ways that define a logical deduction and yet still succeed in allowing people to accomplish goals. This raises important questions about the role of deductive logic within the psychology of thinking.

The next chapter in this book discusses the psychology of decision-making and probability estimation. Many of the cognitive biases that undermine deductive reasoning will also undermine sound decision-making. But, as with deductive reasoning, evidence suggests that many people still make adaptive and smart decisions despite these biases.

CHAPTER 14

How We Decide

In the early part of 2020, schools, companies, governments and individuals had some decisions to make about COVID-19. As the virus was expanding beyond the initial outbreaks in China and Italy, it became clear that it was not going to abate on its own. Most leaders and public health officials agreed that more drastic measures would be needed to slow things down so that the healthcare system was not overwhelmed. An overwhelmed healthcare system is a problem, because neither COVID patients nor other patients will be able to get the help they need. One of the strategies that was considered and implemented in most places was the shutdown or a lockdown. The details might differ from region to region but when a city or country enters a lockdown period, most shopping is closed, malls and theatres are closed, concerts and sporting events are put on hold, schools are closed and/or switch to online instruction. The idea is that if most non-essential businesses are shut down, people will stay home, stay indoors, and stay away from close contact with others. This in turn might slow the spread of the virus. Might. That's the critical point. The outbreak was unlike anything that had ever happened before and so much was uncertain.

It was not the first time that many regions across the globe faced the same major crisis at the same time. The twentieth century saw world wars and pandemics too, but it was the first time in most of our lives that something so big was happening all at once. It was the first time that a global pandemic spread in an era when information and news travelled even faster and more aggressively that the virus itself. And because this virus was so new, so novel, the threat also came with considerable uncertainty. Reflecting back on the early stages of the pandemic, it's like nothing I've ever lived through before. And I suspect that's true for most readers as well.

I watched as different national governments, health officials and pundits weighed the pros and cons of locking down their economy and lifestyle. In each case, a government (local, state/provincial, or national) had to consider several variables, known facts, unknown facts, risks, probabilities and outcomes. How many cases were there in the region? What was the rate of transmission? What was the risk of hospitalisation? What was the risk of death once diagnosed? These were just the first part of the complex equation. Others were more difficult still. How long could the region be shut down? What were the short-term costs of closing businesses? What were the long-term costs to the overall economy? What was the probability of people following guidelines and rules? There were so many things that people needed to consider. People want certainty and to be assured, but this crisis seemed to offer very little of either. There was intense time pressure as well, which limits the amount of time to consider the options and can also change the outcomes.

Some governments chose to pursue lockdowns and shutdowns early on. China, Italy, Canada, Germany, New Zealand and many other countries all chose to pursue widespread shutdowns of their economies as they tried to slow the spread of the virus. Other regions chose to wait and see how things would spread before they took action. The UK initially chose to pursue a strategy that would allow the virus to spread to achieve some level of immunity in the population. Sweden kept as many options open as possible. Other regions, like South Korea, did not lock down or shut down as stringently but focused their efforts instead on quarantining those with the virus and tracing who they had contact with. And finally, some countries like the US and Brazil, adopted unstructured and patchwork approaches that sometimes seemed to undermine their own efforts at controlling the spread.

There is still so much that is unknown about the novel coronavirus. The outcome of many of these decisions may not be known for months or years. But what is clear is that, for many governments and individuals, they never had to make so many decisions with so much at stake and so little prior information to go on. Many of our decisions are informed by the outcome of past decisions. We look to prior successes and failures to guide present decisions. We also try to rely on fast and frugal heuristics when time is at stake (System 1) and slower, more deliberative responding (System 2) when we have more time and the costs might be higher. The COVID-19 pandemic did not fit neatly into any existing prior schema and as I discussed earlier in Chapter 6, there were some notable and disastrous outcomes that resulted from leaders relying on memory-based heuristics, especially early decisions to tell people to get out and keep having fun. Later decisions by state governments in the US to re-open too early may also have been the wrong decision. Individual decisions by people to suspend or not suspend their usual summer leisure activities may have had unintended consequences. The results and reactions to all these decisions will continue to reverberate for years. Decisions in a state of uncertainty, whether big or small, can have unforeseen outcomes. And that can be unsettling.

Decision-making is all about reducing uncertainty, minimising risk, and maximising benefit. I began with the decisions around COVID, but that's an extraordinary case. Many of the decisions we make are trivial, almost unnoticeable. Deciding whether to have toast or a bagel for breakfast is still a decision, but it is one that has little or no uncertainty, very little risk is associated with either outcome, and only a modest benefit. But other decisions are much more serious. When you are deciding on a university major or plan of study, there is uncertainty associated with the outcome. Choosing between engineering and epidemiology carries with it many unknowns. What will the job market be like for engineers and epidemiologists in five years' time? What are the risks associated with each programme? How difficult are the courses? What is the completion rate of students who enter the programme? How likely is it that you will be able to finish near the top of your class versus the middle? Will that matter?

We do not like uncertainty. Animals do not like uncertainty either. Uncertainty makes decisions more difficult and taxes the cognitive system too, by sometimes introducing an unknown number of outcome scenarios. Uncertainty can introduce a state of anxiety as well. For these reasons, most organisms behave in ways that reduce uncertainty and maintain a status quo. After all, the best way to make sure you know what's coming next is just to keep doing the same thing and to maintain as much of the status quo as possible – even if the status quo is not that good. Even a bad situation that you are familiar with might seem preferable to an unknown and uncertain future.

The idea of uncertainty reduction, risk avoidance and the maintenance of the status quo are central to understanding how humans make decisions. If it helps you to decide to keep reading this chapter, I'll try to help by reducing the uncertainty of what's coming next. I'll discuss the steps and stages of decision-making first. Then turn to a discussion of probability, which is critical to understand how we decide things. Then I'll introduce several theories that explain human decision-making in terms of reducing uncertainty and maximising outcome. I'll end with a discussion of how to make the best of uncertain situations and to consider approaches that will help you make good decisions, and to be happy about the decisions you make.

Making a Decision

We make many decisions every day. We decide what to have for breakfast or which route to travel to work. We decide how to allocate time, money and resources. We might decide to stay with a romantic partner or to leave. We decide to stick with a frustrating job, or we decide to leave that job for another. These decisions can be trivial or life-altering. They can be made quickly or with extensive deliberation. They can be right, wrong or neither. We want to reduce uncertainty but it affects everything we do.

Three steps to decide

Decisions that are made with some degree of over-awareness often involve several steps. You might not be aware of each step as you take it, and not every decision involves each step. The first is an identification step, in which you identify the need to make a decision. This might be something as simple as being confronted with an overt decision opportunity, like ordering food at a restaurant, or something more complicated, like deciding how to invest some of your money in Tesla stock. At some point prior, there was no need to make a decision, but at this identification phase, the need for a decision is clear. More importantly, the decision is framed. Framing a decision involves stating the decision in terms of known costs and benefits, or perceived gains and losses. The way in which a decision is framed can alter the way the decision is made. For example, if you are deciding to take a course at university or a training session at your job, it may change the decision if you frame it as a requirement versus framing it as an option. One alternative is framed as something that you have to do; the other is framed as something that you want to do.

A second step is the generation of decision alternatives. For example, if the decision is about where to take a romantic partner on a planned date, you might start to think of the options: movie, club, dinner, golfing, the beach, etc. Like the recognition phase, the generation phase is affected by several factors. Individual factors, such as personal knowledge and experience, can play a role by either constraining or enhancing the alternatives that can be generated. Cognitive factors, such as working memory capacity, can have an effect by reducing how many alternatives can be generated. Environmental factors, such as the amount of time available, can also affect the number of alternatives that can be generated; time pressure reduces the number of alternatives. Making fast decisions is often helped by heuristics and System 1, but as I discussed earlier with COVID-19, the accelerating pace of the pandemic in early 2020 placed additional pressure on decision makers.

The generated alternatives are evaluated in a judgement phase. Judgements are made about probabilities, costs, benefits and the value of alternatives. They can be made about real or perceived risk. In many cases, judgements are susceptible to many of the biases that have been discussed in Chapter 11 and other sections of this book. Both availability and representativeness can affect how alternatives are assessed and evaluated. For example, alternatives that come to mind very quickly might be judged as favourable, but this is a direct result of the availability heuristic. In some cases, alternatives that are highly salient, and thus available in memory, are likely to be brought to mind quickly. This can lead to a bias, and occasionally errors if these alternatives are not optimal. But some of these shortcuts and heuristics are helpful.

What if there are too many options?

In general, we tend to make good decisions. Not exclusively, of course, but many decisions simply do not involve much risk or uncertainty. But there are occasional conflicts. Sometimes, there are simply too many choices. And too many choices can make it harder to decide by taxing your resources when you generate and judge alternatives. When I was a teenager, I did what most teenagers did: I thought about what music to buy. In the 1980s, that meant buying music on a cassette and later a compact disc. Because each purchase was substantial, I spent time in record stores looking at each album, reading reviews and talking to friends. It was fun to decide. It was fun to listen. And I moved these purchases with me from home to university and grad school. The advent of digital music in the 2000s did not seem like a big change, I'd just buy and download a file. But the decisions, though seemingly easier because they were all online, became more difficult because there were more things to choose from and at the same time, they all looked the same. That spiralled out of control as streaming became the main way to consume music. There were suddenly too many choices. As someone whose habit had been shaped by choosing albums, streaming no longer made sense. Companies like Spotify offered all the same albums but also remastered versions, deluxe editions and singles. It really took a lot of the fun out of music for me, and I now rely more on Spotify algorithms and curation to make the decisions for me. With so many decisions, so many options, so many choices, it's just easier to let someone else decide. We used to call that radio, and it was free. Now it's curated playlists and I pay for it. Or think of how many video media choices you might have. At one time, people would have fewer than ten TV channels to choose from and we would select a show from that small set. In the early 2020s, there are so many streaming options (e.g. Netflix, Disney+, Amazon Prime Video, etc.) that it seems like an infinity of options. With that many options, it can be difficult to choose unless you rely on a heuristic or some kind of strategy to limit the options.

Barry Schwartz, a psychologist at Swarthmore College, has written about this in his book, *The Paradox of Choice* (Schwartz, 2004). He notes that too many decision options place a burden on our cognitive system and can undermine happiness and the ability to make good decisions. More options also increase the probability that you will make the wrong choice. Or they can increase the tendency to worry that you will make the wrong choice and *that* can be frustrating. Think of a restaurant menu with pages and pages of different choices. Do you find it difficult to choose? I do. When one of my girls was younger, maybe around seven or eight years old, she used to express frustration when we went out to eat because of all the choices on the menu. The problem is that she liked nearly everything. As a result, she could not decide what she would enjoy the most. At some point, she hit on a strategy to reduce her frustration. She relied on a heuristic and if she could not decide, she would resort to a standard order: A chicken-Caesar wrap, for example, is pretty standard family restaurant and pub fare in Canada. I often do the same thing at certain restaurants with long menus. I find something that I know I will like, and I just order that. That's a strategy that Schwartz calls satisficing. It means choosing something that will work and satisfy a criterion. It may not be the best decision, but it's a good one. A strategy like this also reduces uncertainty by selecting something you are certain to enjoy.

Satisficing has its origins in cognitive science and the pioneering work on human problem solving by Herbert Simon (Newell, Simon & Others, 1972). Simon defined satisficing as setting a criterion or aspiration level, and then searching for the first alternative that is satisfactory according to that criterion. Satisficing is sometimes called a 'good enough' approach because, in many cases, it is a strategy that is geared towards finding an alternative that is good enough, but maybe not the best. It is a strategy that is explicitly suboptimal.

We tend to assume that it is optimal to be optimal. In other words, there is a psychological and behavioural premium on optimality decision-making. This has an intuitive appeal because if we define optimal as 'ensuring the best outcome', it is hard to argue against optimality as the preferred state. But is there any downside to preferring optimality? Consider this scenario. You are in an airport and you have not eaten yet. You have a three-hour flight ahead of you and you need to find something to eat in the next 30 minutes. There are probably hundreds of choices, and you want to get the best meal you can in terms of quality and value. It has to be fast, not too spicy, but also good tasting and healthy. You could stop at each restaurant and vendor and find out what they have to offer and compare this with reviews on social media and review sites. Using this combination of things, you might be able to arrive at the optimal pre-flight meal, but this exhaustive search and evaluation procedure might take a long time, possibly too long given the circumstances.

Instead, you might look at two or three options that are close to you and choose a meal that comes close to satisfying your criteria. The best solution seems to be to set a flexible and modest threshold and choose the first option that seems to meet the basic criteria. The cost of evaluating all of the options to choose the optimal meal is significant, and the benefits are not that great. In addition, there is not much cost to a less than perfect meal in this scenario. In the airport food example, you might decide to get something for less than $15.00 that is not spicy. This can be met with the first sandwich, hamburger or sushi platter that you find.

Understanding Probability

In order to understand the decision-making process, it is useful to have an understanding of how probability works and how people usually assess probability. Many decisions are made with a reference to the probability of an outcome or in the face of uncertainty. When we don't have that information, we can only rely on very general heuristics and are at a risk of making errors. Humans (and to some degree, non-human animals) have a number of ways to track and interpret these probabilities. Jonathan Baron, a psychologist at the University of Pennsylvania, describes three primary ways in which people make sense of probabilities (Baron, 2008). We rely on frequency tracking, knowledge of probability logic and personal theories.

Frequency tracking suggest that humans make probability judgements on the basis of their knowledge of prior frequency events. For example, if you are thinking about the chances that you will catch influenza (seasonal flu) this year, you might base your judgement of the probability on your knowledge of the frequency of catching the flu in the past. If you have never had the flu, you are likely to underestimate the probability. If you got the flu last year, you may overestimate the probability. The true probability may be somewhere in between those extremes. The important thing about frequency theories is that it requires attention in order to encode the event and memory in order to make the judgement. As discussed earlier in Chapter 6, memory can be quite susceptible to bias. We tend to remember highly salient events and base judgements on these memories, which gives rise to the availability heuristic. Availability can lead to bias and decision errors if low frequency information has a strong memory representation because of its salience or recency.

People can use an understating of logical probability as well. This requires knowledge of the actual probability and base rate of a given event. In practice, this can be difficult because probabilities are affected by many factors. However, people can make use of logical theories for what are known as exchangeable events. An exchangeable event is one for which the probability is known and not affected by different forms, different days and different surface features. Standard playing cards are an example of exchangeable events. The probability of drawing an ace of clubs from a standard deck of cards is the same for all standard decks. And it is the same for all people. And it's the same today as it is tomorrow. The probability is exchangeable because it is not affected by these environmental factors. Furthermore, it is not affected by frequency and availability. The likelihood of drawing an ace from a standard deck is the same whether or not you have drawn aces in the past. Even if you have never drawn an ace of clubs from a deck of cards, the probability does not change as a function of your personal memory for the frequency.

Baron points out that purely exchangeable events are difficult and nearly impossible to find. And even on those occasions when truly exchangeable events are being considered, they may still be subject to personal and cognitive biases. For example, if you draw many aces in a row (or no aces) you may experience a shift in expectations as a result of the conflict between your logical probability knowledge and the knowledge gleaned from your own frequency theory. This is known as a gambler's fallacy and I'll have more to say about that later.

Baron also suggests that people make use of personal theories as well. Personal theories can contain information about event frequency and information about logical probability but can also contain additional information. Specifically, personal theories contain information about context, expert knowledge, what should happen, and what you want to happen. The personal view is very flexible because it takes into account the personal beliefs and understandings of the decision-maker. These beliefs can differ among and between people as a function of personal knowledge, and so two people can reasonably be expected to differ in their assessments of probability. Experts and novices would be expected to differ as well. For example, a naïve medical diagnosis made via consultation with a medical website can differ from a diagnosis made by an experienced physician because the physician has specialised knowledge and diagnostic experience. On the other hand, one of the shortcomings of the personal view is that it assumes that people make use of idiosyncratic and irrational information as well. Beliefs about luck, fate, magic and divine intervention can affect our personal theories of probability. These can be difficult to evaluate objectively, but they still affect people's decisions.

How to calculate probability

At its most basic, probability is described as the likelihood of some event occurring in the long run. If something can never happen, the probability is 0.0. If something always happens, the probability of occurrence is 1.0. Most probabilities are therefore somewhere in between 0.0 and 1.0. An event with a .25 probability of occurrence suggests that on 25 per cent of all possible occurrences that event will occur, and on 75 per cent of all possible occurrences that event will not occur. Probability can also be described as the odds of occurrence, such as 1 out of 4, and that means the same thing, but framing it that way can lead to a bias of small number. If I have a 1 in 4 chance of winning, I am more likely to expect to win every four times. As we'll see, that may not be the case.

To calculate the probability of occurrence according to the simplest logical theories, the number of desired outcomes is divided by the number of possible outcomes. For a simple coin toss, the number of possible outcomes is two: heads or tails. The probability of getting a head on one coin toss is 1 divided by 2, which equals .5. This corresponds with our intuitions that a fair coin has a .5 probability of heads and a .5 probability of tails on any toss. Practically, that means that if you toss the coin several times in a row, you tend to expect some heads and some tails. You don't expect it always to be an even distribution of heads and tails, but you expect that over many tosses (or an infinite number of tosses) the frequency of heads and tails should balance. We also assume that these probabilities hold up over the long run but allow for variation in small numbers of samples.

Combining probabilities

When calculating the probability of multiple events, the individual probabilities need to be combined. That is, they are either be multiplied or added, and that's where we start to see some problems. The easiest way to remember is that, when the probability is combined with an 'and', you multiply to combine. When they are combined with an 'or', you add.

For example, to calculate the probability in a coin toss of obtaining two heads in a row (head AND head), you would take the probability of one head (.5) and multiply it by the probability of another head (.5). The probability of two heads in a row is .25, and three heads in a row is .125, and so on. This means that the probability of several heads in a row is lower by multiplicative factor. When calculating the probability of getting two heads or two tails (head/head OR tail/tail), you add the probabilities. So, the probability of two heads is .25 and the probability of two tails is .25, which results in .5. All four combinations (head/head OR tail/tail OR head/tail OR tail/head) adds up to 1.0 because those are all the possible options for two coin tosses in a row. These rules both assume independence. That means that the result of the first coin toss has no effect on the result of the second coin toss. Although the probability of two heads in a row is .25, the probability of each head alone is still .5. This means that even if you tossed a coin twenty times in a row and obtained twenty heads in a row, the independent probability of the twenty-first toss is still .5 for heads. These events are entirely independent.

Gambler's fallacy

Confusing an independence with what we believe to be representative of randomness is known as the gambler's fallacy. This happens when your personal theories and beliefs intrude on logical theories. Sometimes the gambler's fallacy arises from the representativeness heuristic. Suppose you flip a coin ten times and each time you record if it's head or tails. Now consider these three following sequences of head (H) and tails (T) that could result from the ten coin flips. Note that, according to our description of multiplicative probability, all three are equally likely.

Example 1: H–T–T–H–T–H–H–T–H–T
Example 2: H–T–H–T–H–T–H–T–H–T
Example 3: H–H–H–H–H–H–H–H–H–H

Do they seem the same? Although each sequence has a 0.000976 probability of occurring, Example 1 might seem like the most representative example of randomness and Example 3 to not be very representative of randomness because it's all heads. What would happen if you were asked to bet on the outcome of the next coin flip? That is, what would you expect the eleventh coin flip to be? If you're like most people, you would probably have no preference for Example 1 (50/50 head or tails) and you might not even have a strong feeling about Example 2, but if asked to gamble on the next coin flip for Example 3, you might strongly expect there to be a tails (T). After ten heads, a tail seems due to come up.

That's the gambler's fallacy. This is a systematic overestimation of the probability of tails on the eleventh coin toss after a sequence of ten heads. Because the coin toss is known to have a true probability of .5 heads and .5 tails, a sequence of ten heads seems unnatural and non-random, even though it is a random occurrence. If people estimate the probability of obtaining tails on the eleventh flip as greater than .5, they are falling prey to this gambler's fallacy. It is a difficult fallacy to overcome. Even if you know that the coin toss is independent and always has a .5 probability of coming up heads and a .5 probability of coming up tails, most of us would feel very strongly that the tails toss is due after ten heads in a row.

It is easy to see how independence and the multiplication rule work for simple events like coin tosses, but these effects become stronger in more complicated and semantically rich examples because our knowledge can override probability. We sometimes ignore it even when it's given. A frequently cited example comes from Kahneman and Tversky (Tversky & Kahneman, 1983). Subjects were shown a description of a person and were then asked to indicate the probability that that person belonged to one or more groups. The most well-known example is the 'Linda' example.

> Linda is 31 years old, outspoken and bright. She majored in Philosophy, and as a student she was concerned with issues of social justice and discrimination. She also participated in many demonstrations.

After reading the description, subjects were then asked to rate the likelihood that Linda was a member of several groups.

- Linda is a teacher in elementary school.

- Linda works in a bookstore and takes Yoga classes.

- Linda is active in the feminist movement.

- Linda is a psychiatric social worker.

- Linda is a member of the League of Women Voters.

- Linda is a bank teller.

- Linda is an insurance salesperson.

- Linda is a bank teller and is active in the feminist movement.

They found that people rated the likelihood that she was a feminist as being high because the description fits a stereotype or category of feminist.[40] They also rated the likelihood that she was a bank teller as relatively low. There is no reason that Linda cannot be a bank teller, but there is nothing in her description that is strongly indicative of bank teller either. Furthermore, bank teller is a broad category with less clear attributes. The key result is that subjects rated the likelihood that she was a bank teller *and* feminist as being greater than the likelihood that she was a bank teller alone. Logically, it is not possible for the likelihood of a conjunction of two categories to be higher than the likelihood of any one of the single categories. In this case, the strong semantic connection between this description and a stereotype of the feminist movement causes this error to occur. Subjects ignore the conjunction and focus on the fact that Linda is representative of being a feminist. In other words, the representativeness heuristic is stronger than our sense of logical probability.

Cumulative risks

People also make errors in how they apply the adding rule to cumulative risks. As an example, imagine the likelihood of having a car accident while driving to work. On any given day, the probability is low. But over several years of driving daily, the cumulative risks increase. This is because the cumulative risks in this case are calculated via adding. That is, over ten years, an accident on Day 1 or Day 2 or Day 3, etc. To calculate cumulative risks of being in a car accident on any day out of several years' worth of driving,

---

40 A few comments. This study was carried out in the 1980s though most of us might still activate a stereotype for Linda that is close to what they had in mind. We might think of 'feminist' differently than people did in 1980, but many of the features of someone who is active in social justice would be comparable. Second, the example of 'bank teller' might not make as much sense as it did when the study was originally conducted. Although most people can get cash from their bank from a cash machine (or just pay with a debit card or even your phone), it used to be that the main way to get cash was at a bank. The people who worked at the front counter are called 'tellers' and they would be able to handle basic transactions. There is no specific type of person who would work as a teller. They still exist, but most people do their everyday banking online, on their phone or at a cash machine, an ATM (an Automated Teller Machine). Third, the League of Women Voters is less familiar and is a nonpartisan American organisation originally for women only, that was formed after women won the right to vote.

you would essentially be adding the probability of the accident today with the probability of an accident tomorrow with the probability of an accident the day after that. And the probability changes and fluctuates. A very small risk of an accident on any one given day adds up to a larger cumulative risk over the long run.

As another example, consider the decision that many people make to send or receive a text or DM from their smartphone while driving. It's against the law in most places. There are fines. Most people know that this is a distraction. Given that the risk is known, and the fines are steep, it is surprising and discouraging that people still use their smartphones while driving. One possibility is that people do not truly understand the cumulative risk. This is especially true if drivers are basing their decisions on a frequency theory. Unless you have had an accident or a fine, your own perspective and interpretation of the probability is that every time you use your smartphone you do not have an accident. Even drivers who have been using a smartphone while driving for years may have never had an accident. This does not mean that the smartphone is not interfering; it just means that the interference did not result in an accident. And so, it is not unreasonable to imagine that a short-term decision based on frequency knowledge of the risk associated with using a smartphone would cause the driver to believe that the smartphone is not a danger, even if that belief conflicts with what the driver knows about the real risks. With no prior personal connection between smartphone use in accidents, most people downplay the risk. While this is sensible in the short-term, it is clearly dangerous in the long-term.

Calculating probability over the long run, whether it is a normative value or a cumulative risk, is often difficult for people. In many ways, it is counterintuitive to make judgements about things that happen in the long run. In the long run, a coin may have a .5 probability of coming up heads. But this assumes an infinite number of coin flips. A large number of coin flips will tend to balance out, but a small number may not. The problem arises when we make decisions to maximise short-term gains rather than long-term probabilities. From an evolutionary standpoint, this makes sense. Short-term probabilities have some advantages. An organism needs to eat today. It may not be able to consider what is going to happen two months, two years or two decades in the future.

Base rates

The rate of occurrence for an event (i.e., the long run probability) is known as its base rate. For many exchangeable events, like cards, the base rates may be known. For most events, though, the base rates may not be known explicitly, or they may be estimated via knowledge about the frequency of occurrence. For example, it is nearly impossible to know the base rate of occurrence for the likelihood that the bus you take will not be on time. Or the base rate of a vehicle accident. Furthermore, even if you do have some knowledge about base rates, (e.g., the likelihood that the bus you take to work will break down), it is probably a very low probability. A very low base rate means that the event does not happen very often, and so you are likely to ignore this base rate. That's natural.

Given the difficulties in understanding and using base rates, it should not be surprising that people often ignore them even when they are known and when using them would be useful. This is known as base rate neglect. A striking example is seen in medical tests and diagnoses. A series of studies by Gerd Gigerenzer investigated how physicians, nurses and even non-medical subjects arrived at decisions about the efficacy of the medical test (Gigerenzer, Gaissmaier, Kurz-Milcke, Schwartz & Woloshin, 2007). Most of the experiments presented subjects with scenarios in which a given disease has some established base rate. That is, an established rate of occurrence in the general population. Subjects were then given information about tests.

For example, consider a serious disease for which early detection can be very beneficial. This disease has a base rate of 1 per cent. This means that 100 in every 10,000 will eventually contract the disease. Early detection requires the administration of a diagnostic exam. The test has a very good hit rate. If the disease is present, 98 per cent of the time the test will indicate a positive result. If the disease is not present, only 1 per cent of the time will the test indicate a positive result. In other words, the test has a high hit rate and a very low false positive rate. On the surface, this would appear to be a very good test that can reliably distinguish people who have the disease from those who do not. Given this information, and given a positive test result, how would you rate the likelihood of a patient actually having the disease?

In order to understand how to arrive at this decision, it is necessary to explain first how to calculate this probability. I've set up table with four probability cells and I'm using a population of 10,000 (though you can scale this up or down). The 1 per cent base rate, remember, assumes that 100 in every 10,000 people will have this disease. In the 'Disease is Present' column, ninety-eight of those hundred people with the disease will show up as a positive test result. And two of those hundred people will show up as a negative test result, despite having the disease. Pretty good so far, only missed two.

But the problem emerges in the people who do not have the disease. There are 9,900 of them and we're testing them all. In the 'Disease is Absent' column, the 1 per cent false alarm rate means that 99 people out of the 9,900 that do not have the disease will still show a positive test result, whereas 9,801 will show a negative test result when no disease is present. So, it's a good test, but because there are so many people without the disease, it means we're misdiagnosing ninety-nine of them. Still think this is a good test? It gets worse.

We need to decide what a positive test result really means. After all, we don't actually know who has the disease: that's why we're testing them. To calculate the probability of a person having the disease *given* a positive test result, *P(disease|positive test)*, we need to take into about the number of people with the disease who test positive and divide by the number of people who test positive in general. Think of this as a scaled-up version of *P(head) = heads/head+tails*. That can be done with the following formula (which is simplified from Bayes theorem.):

$$P\ (disease|positive\ result) = \frac{people\ with\ disease\ who\ test\ positive}{people\ test\ postive}$$

Using the values in the table below, we have:

$$P\ (disease|positive\ result) = \frac{98}{198} = 0.497$$

Assuming a base rate of 1 per cent, which works out to be 100 in every 10,000, the probability of a person having the disease given a positive test result is .497. Now that test, which seemed so promising, is no better than a coin toss. Even a very diagnostic test can result in a low conditional probability if the base rate is low.

Table: Probabilities for a hypothetical disease and diagnostic test

Test Result	Disease		
	Disease is Present	Disease is Absent	Totals
Positive Test Result	98	99	197
Negative Test Result	2	9,801	9,803
Totals	100	9,900	10,000

The critical information that is often ignored is the number of people with false positives. Although the false positive rate is low, the base rate is low as well. That means that most people actually don't have the disease. If even a small percentage of those people without the disease show a false positive, the *absolute number* of false positive cases is higher than desirable. In other words, despite a test with a very high hit rate and a very low false alarm rate, a low base rate results in a low probability of the person with a positive test actually having the disease. In many of Gigerenzer's studies, even experienced physicians overestimated the probability of a positive test result indicating the actual presence of a disease.

This raises an interesting question. If even physicians make this error, how do they keep from letting this error affect their decisions in real, clinical settings? One way is to take steps to increase the base rate so that a positive result on a test is more diagnostic. This is done by only ordering tests for individuals who are more likely to have the disease. In other words, you don't order a cancer screening test for everyone, but your order one for patients with other symptoms and existing risk factors. This effectively increases the base rate by applying the test to people who are already more likely to have the disease. Testing everyone, even with a test that has a high hit rate and low false positive rate is not a good strategy when the incidence of the disease is low. But by testing people who are already more likely to have the disease, you ensure that the outcome of the test is more useful and more informative.

Rational and Not-So-Rational Decision-Making

Decision-making involves a combination of knowledge about outcomes, costs, benefits and probabilities. The previous section covered understanding probability and the errors that may result. This section covers several theoretical approaches that attempt to explain and understand how people make decisions.

A rational approach to decision-making sets a normative standard from which we can investigate deviations. In other words, many of the characteristics and aspects of this model are rooted in economic theory and may not describe the cognitive processes behind decision-making as well as some of the other theoretical approaches we will take. On the other hand, this rational approach describes how one might arrive at an optimal decision. The degree to which people deviate from this optimal decision can be understood in the context of other theoretical approaches, memory, knowledge and cognitive limitations.

A fundamental aspect to the rational approach is the assumption that people can make optimal decisions. In order to make an optimal decision, it is assumed that people weigh the alternatives, they set an expected value or determine the expected utility for all of the alternatives, and then they proceed to choose the most valuable alternative in the long run. That is, the optimal decision is the one that maximises the expected value/utility.

The expected value can be thought of as a physical, monetary or psychological value that is attached to a given outcome. An expected value is calculated by combining what is known about the costs and benefits with the probability of achieving a desired outcome. The formula below shows a straightforward way to calculate an expected value:

$$EV = (Value_{Gain} * P (desired\ outcome)) - (Value_{Loss} * P(non\ desired\ outcome))$$

In this formula, expected value (EV) is a function of the gain and the probability of achieving that gain minus the costs and the probability of incurring those costs. For example, consider a very straightforward set of monetary choices. One option (1) allows you to gain \$40 with a probability of .2 or else nothing. The other option (2) allows you to gain \$35 with a probability of .25 or else nothing. The first choice has a higher win but a lower probability. In order to decide which of these is the best option in the long run, the values can be inserted into the expected value formula as follows:

Option 1: $EV = (\$40 * .20) - (0 * .80) = \$8.00$
Option 2: $EV = (\$35 * .25) - (0 * .75) = \$8.75$

Despite having the higher pay out, Option 1 actually has a lower expected value over the long run. This is because of the way gain and probability combine. This also means that different factors can affect the long-run expected value. In cases where the outcome has some additional utility beyond the actual value, it can affect choice by increasing the value of a win or decreasing the cost for a loss. For example, many people engage in gambling and buying lottery tickets, despite the negative expected value for these events. Most people agree that there is an additional psychological utility that comes from a casino game. Gambling in a casino might be fun, it might be an enjoyable social activity, or it might be part of a vacation. This would essentially minimise the impact of a loss.

Framing and loss aversion

As previously discussed, the rational approach is very effective at describing the optimal decision-making patterns, but very often fails to describe how people actually make decisions. People often make decisions that are in opposition to expected values and optimality. A fairly straightforward example, using the same general framework as the simple gambling option earlier, is known as a certainty effect. All things being equal, humans and many other non-human animals resist uncertainty. Uncertainty creates an unwelcome state for any organism. But sometimes, the context around a decision can activate memories, known as 'frames', to make some options seem more certain that they are. Originally described by Tversky and Kahneman (Tversky & Kahneman, 1981), the framing effect is an illustration about how the context and semantics of a decision can affect which options seem preferable. The most well-known example consists of a short description of a scenario and two choices. Consider the following statements:

Premise: Imagine that the US is preparing for the outbreak of an unusual disease, which is expected to kill six hundred people. Two alternative programmes to combat the disease have been proposed. Assume that the exact scientific estimate of the consequences of the programmes are as follows:

> **Programme A.** If Programme A is adopted, two hundred people will be saved.
> **Programme B.** If Programme B is adopted, there is 1/3

probability that six hundred people will be saved, and 2/3 probability that no people will be saved.

In this scenario, most participants (72 per cent) choose Programme A. Although it is implicit in that statement that two hundred people will be saved and therefore four hundred people will not, the framing is in terms of *lives saved*. Both programmes talk about lives being saved. When framed as a saving or as a gain, people generally display risk averse behaviour and prefer a certain outcome.

However, imagine the same scenario but framed as the number of people who will die:

> **Programme C.** If Programme C is adopted four hundred people will die.
> **Programme D.** If Programme D is adopted there is 1/3 probability that nobody will die, and 2/3 probability that six hundred people will die.

In this scenario, most participants (78 per cent) choose option D. Notice that the numbers are equivalent between Programmes A and C and B and D across both scenarios. This second scenario is framed as a loss (i.e., people dying). When a decision is framed as a loss, people generally display loss aversion and are more willing to choose the riskier alternative. This framing example is interesting because it pits loss aversion against risk aversion. Although people generally like to avoid risk and loss, avoiding loss is paramount. Loss aversion looms large in human behaviour. Curiously, both may stem from a desire to avoid uncertainty. People tend to avoid risk as a way to reduce uncertainty, but people also tend to avoid loss as a way to avoid uncertainty and protect the status quo.

Nudges

Many retail environments and stores take advantage of our preference for certainty, loss aversion and risk aversion in the way they advertise. For example, every summer I go with my kids to purchase new sporting equipment for baseball, and in the winter for ice sports. One year, we went to a local sporting goods store which had a sign advertising that everything was on sale in the store, up to 50 per cent off. My older daughter, who was about ten at the time, said, 'It is a good thing we decided to come today, they're having a big sale.' Which is a good point. It must have seemed like great timing. I hated to burst her bubble, but I pointed out that the store always has that sign up. What is listed as the 'sale price' is just the actual price. This is seen in other retail environments where the tag may have a 'suggested price' that the object is never sold for. These pricing strategies present a framing effect whereby the advertised price appears to be a bargain.

The work of Richard Thaler (Thaler, 1985) has examined this idea more systematically. For example, consider the following descriptions of two gas (petrol) stations with different pricing strategies. In one of Thaler's studies, participants were presented with the following scenario:

> Driving down the road, you notice your car is running low on gasoline, and you see two service stations, both advertising gasoline. Station A's price is $1.00 per litre; station B's price is $0.95 per litre. Station A's sign also announces, '5 cents/litre discount for cash!' Station B's sign announces, '5 cents/litre surcharge for credit cards.' All other factors being equal (e.g., cleanliness of the stations, whether you like the brand of gasoline carried, number of cars waiting at each), to which station would you choose to go?

With no additional context, subjects tended to report a preference for Station A. The costs were identical regardless of whether subjects were asked to pay with cash or with credit. However, the first station's policy was framed as a discount for paying cash. This is interpreted as a more favourable pricing strategy and so that station was preferred.

Thaler's work, which eventually culminated in him winning the 2017 Nobel Prize in Economics, is outlined in his recent book *Nudge: Improving Decisions About Health, Wealth, and Happiness* (Thaler & Sunstein, 2009), in which he and his co-author Cass Sunstein argue that governments and companies can help to improve decision-making and ultimately improve society by adopting policies that will nudge people in the optimal direction. They define a nudge as:

> *A nudge, as we will use the term, is any aspect of the choice architecture that alters people's behavior in a predictable way without forbidding any options or significantly changing their economic incentives. To count as a mere nudge, the intervention must be easy and cheap to avoid. Nudges are not mandates. Putting fruit at eye level counts as a nudge. Banning junk food does not.*

The idea behind nudge theory is that we can use our biases to our advantage, to make better decisions.

Loss aversion

Several of the preceding examples dealt with the phenomenon of loss aversion. Loss aversion occurs because the psychological value assigned to giving up something or losing something is greater than the corresponding psychological value associated with obtaining that same object. We tend to hold on to the things that we already have. We tend to prefer current circumstances. Many people continue to use a favourite pen, to keep old books that they like, or to keep a favourite mug. More consequentially, many people stay in relationships that are not ideal for fear of losing what they have. As humans, our decisions are often ruled by loss aversion. The tendency to avoid situations that might lead to loss can lead to a bias known as the status quo bias in which we prefer situations that maintain the status quo. We do this to avoid loss. This status quo bias can affect real world decisions in many surprising ways.

Loss aversion and the status quo bias come in several forms. In the earlier examples on framing, loss aversion came about when participants were likely to prefer riskier alternatives rather than face loss of life in the disease scenario. On a more personal scale, loss aversion associated with small objects is often referred to as the endowment effect. A study by Daniel Kahneman (Kahneman, Knetsch & Thaler, 1991) investigated this effect with university undergraduates. Participants were divided into groups. One group, the Sellers, were given university bookstore coffee mugs and were asked whether they would be willing to sell the mugs at each of a series of prices ranging from $0.25 to $9.25. Buyers were asked whether they would be willing to buy a mug at the same set of prices. Choosers were not given a mug and were asked to choose, for each of the prices, between receiving a mug and receiving that amount of money. In other words, Choosers did not have the mug, and their decision was how much they were willing to pay. It is important to note that everyone ended the study with the same amount. But what is interesting is that Sellers priced the mug at a point nearly twice that of the Choosers. The mere act of possessing the mug endowed it with greater value.

Status quo also shows up in the sunk cost bias, an effect that is also referred to as entrapment. This can be a very effectively used technique in a sales environment. Essentially, people are unwilling to abandon an undesirable status quo primarily because they have already invested time or money. As an example, imagine that you and your friends go to see a movie. In Canada, where I live, the average movie price is somewhere between $12 and $15. Once you have paid your $12, you expect to see a good movie, but what would you do if the movie was truly awful? Would you stay and watch to the end, or would you get up and leave? When I ask this question in a class, most students indicate that they would stay. The reason that is most commonly given is that they already paid for it. The reasoning is that you have paid money (i.e. the sunk cost) for the movie, it is already bad, but leaving the movie will not make it any less bad. And there is always a chance, however remote, that it might get better. You would miss that chance if you left the movie. In short, once you have sunk some cost into something, you desire to see that sunk cost realised.

This effect also arises in other scenarios. If you have ever been on hold with a utility or a telecommunications provider seeking technical or billing support, you will have noticed that it is a common experience to be told something along the lines of 'Your call is important to us, please stay on the line and your call will be answered in the order that it was received'. This can be very frustrating. There is uncertainty with respect to how soon your call will be taken. When I have been in this situation, I often feel that the longer I wait, the more resistant I am to hanging up. I am unwilling to give up sunk time. This behaviour deviates from the optimal model because the cost has already been paid.

Prospect Theory

The biases discussed above suggest that people often make decisions which deviate from rationality. That does not mean that they are bad decisions. And it does not mean that people are making decisions poorly. It just means that there are psychological reasons behind suboptimal decision-making. Kahneman and Tversky proposed an alternative to the standard economic/rational model referred to as prospect theory (Kahneman & Tversky, 1979). Prospect theory suggests that people make decisions according to psychological prospects. In addition, this theory takes into account the difficulty that most people have in correctly judging probability and likelihood. Prospect theory assumes that objective probability can be replaced by psychological probability or beliefs. A key aspect of prospect theory is that loss aversion and risk aversion are primary motivators. Loss aversion looms especially large in this theoretical approach.

Prospect theory is best summarised in the graph shown in Figure 14.1. It shows prospect theory's value function. The x-axis shows losses and gains

Prospect Theory

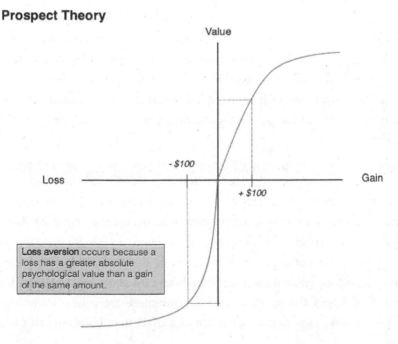

Value

- $100

Loss

Gain

+ $100

Loss aversion occurs because a loss has a greater absolute psychological value than a gain of the same amount.

Figure 14.1: This graph represents the value function for prospect theory. The y-axis shows the psychological value associated with losses and gains and the real value of losses and gains is shown on the x-axis. In prospect theory, the loss curve is steeper than the gain curve.

in terms of real value. The y-axis shows the psychological impact of those losses and gains. There are several things to note about this value function. First, both the loss and the gain curves are concave. That is, there is not a linear relationship between actual gain and the psychological impact of that gain. As shown in the graph, there might be some psychological value placed on gaining $100, but the gain of $200 (twice as much in real dollars) may not be psychologically twice as desirable. According to this theory, the value that you place on gains will eventually reach an asymptote, or a point where the increase in value approaches zero but never reaches it. This is often referred to as 'diminishing returns'.

A second thing to note about prospect theory's value function is that the curves are asymmetrical. The loss curve is steeper than the gain curve. This reflects the loss aversions and general finding that people sometimes value the status quo more than the prospect of a gain. In terms of *absolute value*, a gain of $100 is not worth the same as a loss of $100. As much as we may appreciate an extra $100, we are likely to take steps to avoid a loss of $100. *Losses loom larger than gains* according to prospect theory.

This general approach accounts for many of the findings that show people's deviations from optimality. Kahneman and Tversky argue that prospect theory more accurately describes the psychological process behind human decision-making. We behave sub-optimally because we value the status quo, we seek to avoid and minimise loss, and we seek to reduce uncertainty. These psychological drives may come in conflict with optimality.

Knowledge Affects How You Decide

The rational approach does not rely on personal conceptual knowledge or semantic memory, but rather assumes that decisions are made in accordance with calculated expected values that arise from an understanding of probability and the costs and benefits of each outcome. Prospect theory takes into account personal and cognitive biases by assigning a role for the aversion of risk and the aversion of loss, but there are other, sometimes more idiosyncratic, factors that influence and affect people's ability to make decisions.

In many cases, one of the influencing factors in decision-making is the ability to provide a reason for the decision. Given several options, it is likely that the most attractive decision is the one that has the best justification, even if it does not have a better possible outcome. For example, selecting a menu item at a restaurant because it conforms with a set of beliefs about what you like (seafood) allows you to give a reason: 'I ordered the shrimp risotto because I like shrimp'. This is a great reason to order the dish but does not necessarily reflect a rational analysis. To some degree a prospect analysis accounts for this, if the person wishes to avoid the risks associated with other dishes. But if the diner in question has never eaten risotto before, this decision suggests that risk (risotto) is mitigated by having a good reason to order the dish (I like shrimp).

People often make a decision to avoid experiencing regret for a decision that did not result in a desirable outcome. This is a form of loss aversion. When I teach this topic, I often propose the following example to students. Imagine that everyone in class receives a ticket for a draw worth $100 (this is a thought experiment, not as an actual draw). There is no cost. The tickets are free. Because it is a draw, every ticket has the same probability of winning. No ticket can be considered a winning ticket until the draw has been made. So, when you receive the ticket, your ticket has the same probability of winning as every other ticket in the class. I then ask students if they would consider trading the ticket with the student beside them. Most students indicate that in this scenario, they would be unwilling to trade. The reason is because there is no advantage in trading with someone else because tickets have an equal probability of being winners. However, most students also indicate that if they had traded, and the person next to them won, they would regret having traded. In other words, people avoid the trade in order to avoid a feeling of regret. It would be difficult to avoid feeling regret and to avoid feeling like you had traded away a winning ticket, even though none of the tickets is a winner until the draw is made.

Summary

Decisions underlie many of the important outcomes of cognition. In previous chapters, constructs like memory, concepts and induction reflected internal states and behaviour. With decision-making, these internal states interact with external outcomes. Unlike many of the kinds of thinking that were discussed in earlier chapters, there are real consequences to making good and bad decisions.

In Chapter 10, we discussed the interaction between language and thought. One of the points that I tried to make was that the way in which you describe something linguistically can have an effect on how you think about it. In the current chapter, this was made very clear with respect to framing effects. Decisions that are framed as a loss produce different expectations and outcomes compared to decisions that are framed as a gain. In the most striking example, the terms and outcomes were identical; the only thing that differed was the linguistic and semantic content.

Although most decisions that people make, and most of the decisions that you have made, are straightforward and likely to be correct (or close enough), there are many times when bad decisions are made. There are times when you make a mistake. There are times when you select the wrong course of action. Or there are times when you select one of several uncertain outcomes and things don't go as planned. The research and the ideas discussed in this chapter should make it clear that decision-making is at once a fast and seemingly effortless process, and also a process fraught with potential errors and biases. These biases are the result of heuristics designed to help us. The cognitive biases and heuristics seem to be at play in so many different scenarios and in so many ways, but they may not always be a source of error, but rather of fast and efficient decision-making.

CHAPTER 15

How to Think

In 2020, many people learned or relearned how to work from home. The novel coronavirus that caused COVID-19 also caused a shift in how a lot of people worked. Across the world, teachers, tech workers, knowledge workers, people in finance and people in business began working from home and holding meetings on video meeting platforms like Zoom, Skype or Microsoft Teams. For many of us in this new category, it represented a significant shift in how we did our work, even though the content of much of the work stayed the same. As a professor and researcher, I was one of these people.

Video meetings have been around for a while in academia, but the near total reliance on them was unprecedented. As I discussed earlier in the chapters on memory and induction, knowledge of the past guides our behaviour in new situations. But for this, I had few prior memories available to guide me. What I did have to guide me were my usual routines, like weekly lab meetings and weekly advisory meetings with my students. So that's how I began to structure my online day. It was similar to my pre-pandemic workday but using video meetings in place of face-to-face meetings. I began to teach online using Zoom for student meetings and I began recording video lectures. I met weekly with my graduate students online on Zoom. We held our research lab meetings on Zoom. We had department meetings on Zoom. There were Zoom PhD exams. There were formal Zoom talks and casual Zoom coffee breaks. Some research groups would hold Zoom happy hours. Even academic conferences, which have long been a way for academics, researchers, students and scientists from different locations to come together, switched to online formats. Soon, I was doing all my work – teaching, research, committee work and everything else – from the same screen, on the same computer, in the same room.

Although a lot of my work was able to be carried out easily at home and online, I began to notice some small changes in my own ability. I was becoming more forgetful. I was making more simple memory errors. For example, I might talk with one student for 10 minutes about the wrong project. Or I might confuse one meeting for another. A lot of these mistakes were memory errors that were about the source of the memory. I remembered the wrong source or the wrong event. I wrote about these kinds of misattribution errors in Chapter 6 and they occur when you remember having read something or seen something, but you are not sure where, or you confuse the source of the memory. I seemed to be making more source memory errors than I used to. I felt like I was more like the stereotype of the 'absent minded professor' than I used to be.

Then I realised a possible source of the problem: Everything looked the same. I was looking at the same screen on the same computer in the same room for everything. This was not typical. For my entire career as an academic, I've always had different places for different activities. I would lecture in a lecture hall or classroom. I would hold seminars in a small discussion room. I would meet with students in my office. I would meet with colleagues at the café on campus. Committee meetings were usually held in meeting rooms and board rooms. I would work on data analyses in my office. I would usually write at home or sometimes in a local café. Different places for different tasks. But now, all the work was in one place. Teaching, research, writing and advising were all online. And worse, it all looked the same. It was all on the same screen, on Zoom, and in my home office. I no longer had the variety of space, time, location and context to create a varied set of memory cues.

Why do I bring this up? Because as I noticed the problem, I also thought about my own understanding of cognitive psychology and thought about why I might be making these simple mistakes. As I wrote in Chapter 6, memory is flexible, and it depends on spreading activation to activate similar memories from predictive cues. In some cases, local context can be a strong and helpful memory cue. If you encode some information in one context, you will often remember that information better in the same context. Memory retrieval depends on a connection between the cues that were present at encoding and the cues that were present at retrieval. This is how we know how to adjust our behaviour in different contexts.

We react to locations all the time. When you walk into a restaurant or diner, you probably adjust your behaviour. If you are returning to a restaurant that you were at years ago, you will remember having been there before. Students behave differently in class than out of class. Being in the classroom activates memories from being in classrooms and students adjust accordingly. They might remember what was discussed last week in the same class as soon as they step into the room. You probably do the same kind of thing even if you are no longer a student. You might not think as much about things at home when you are at the office and might not think about your office when you are at home. I wrote about an example in Chapter 8 when I described how my daughter remembered the car on the lift when we went back to the garage. She experienced an event, got on with her day, then re-experienced the event when she saw the same place. Contextual cues help. Being in a specific place helps you remember things that you associate with that place. This is all part of our natural tendency to remember things where and when they are likely to matter most.

But it seemed like this natural tendency was working against me. Each day began at the desk in my home office. Each day I was in the same location when I taught, wrote, met and carried out analyses. But this was also the same location where I read the news, caught up on Twitter and ordered groceries online in the evenings. What I noticed in my new forgetfulness was that I was experiencing some interference. Everything was starting to look the same. The contextual cues that would normally be a helpful reminder of what I was doing were no longer working as memory cues because they were the same cues for everything and everyone. In fact, they were probably having an interfering effect because the location and context information (my office, my desk, my computer screen and Zoom) were the same for so many different meetings that it increased the probability that I would make a source error confusion. When everything looks the same, context is no longer a helpful memory cue. If you work, meet, read, write, shop and casually read the news in exactly the same place, the likelihood that you will make an error of confusion is increased.

This is not an easy problem to solve, of course, because as long as COVID is ascendant, I will still have to work from home. But thanks to my understanding of memory and cognitive psychology, at least I have some understanding of what is happening and why it is happening. And with that information, I might be able to change my work environment to help. For example, a simple fix is to vary the approach to video meetings. It might help to change platforms in a consistent way, say by meeting with one working group on MS Teams and another on Zoom. It's not as strong of a difference as meeting in different rooms, but it's still a change of venue. Another way to accomplish the same goal is simply to change the appearance of your computer each time you meet with a person, using different backgrounds for different people. These seem like very small things and they might not fix the problem entirely, but they could help. More importantly, they are recommendations that come directly from our understanding of how the mind works. If you tried one of these suggestions, like using a different video platform for different meetings, you would be using a psychological theory to make a prediction and then test that prediction with an experiment (on your own behaviour). And that is what I hope you take from the ideas in this book.

Thinking and Cognitive Psychology in Everyday Context

Now that you have gained some understanding of what cognitive psychology is and how it works, you will see examples in your own daily experiences. For example, you might notice the ways in which your attention switches between two tasks and that there is always a short lag in processing when you switch. If you understand what is happening and why it happens you may recognise the problem and will be able to apply insights from cognitive psychology to help you avoid the problem. In doing so, you might adjust your behaviour to reduce the lag by learning to avoid the switch in the first place. Does seeing your smartphone on the desk act as a visual stimulus that pulls your attention? That's what our discussion earlier about visual attention would suggest. One solution might be to keep the phone out of view. Without it there to act as a visual cue, you might be able to concentrate on other things just a bit longer.

Consider another example. Perhaps you notice that you tend to elaborate and fill in detail from memory. Maybe you enjoy telling a story and you add in some slightly exaggerated details to make it more interesting. Or maybe you elaborate on the story to make it more memorable for listeners. Well, given what you know about memory and about the tendency to elaborate, what do you think will be the result in the long run? You will probably remember these elaborated stories better than you would otherwise, but you will also probably remember the elaborations right along with the original memory as they become part of the same event. And this will make it harder for you to distinguish the actual events from the elaborated events.

There are countless examples. Think about each of these questions below. Think about whether or not these apply to you and think about how these examples might be explained and avoided by using the insights from this book.

- Do you find yourself relying on stereotypes when you are reasoning or making decisions?

- Do you make the same mistake each time you try to solve a similar problem?

- Do you ever have trouble looking past someone's appearance because they remind you of someone else?

- Do you have trouble remembering the same simple things each day?

- Do you remember useless things like an old advertising jingle and wonder why it's still active?

For each of these and many other examples, you can find the explanations in cognitive psychology. You can find the answers in many of the chapters in this book. If not directly, then indirectly through a better understanding of the thought process. In my opinion, the best way to learn how to think better and more effectively is to be aware of how mistakes sometimes happen. The best way to notice mistakes and errors in thinking and judgement is to know more about thinking in general. I think an understanding of cognition, cognitive psychology and the brain is useful and helpful for all of us.

How to Think About Your Own Thinking

The title of this chapter, 'How to Think', might suggest that I'm going to tell you how you should think. That would be a reasonable linguistic inference. In doing so, I am not suggesting what to think about. I am not suggesting there's only one way to think. I am suggesting that cognitive psychology can help you understand thinking. From that understanding comes the solution to knowing how to think.

I can't really suggest what you should think about. I can't suggest which past experiences you can rely on. I can only suggest that you will rely on the past experiences that are similar to the present. We all have different experiences. We all have different backgrounds. We speak different languages and some of us speak several different languages. Our memories, our experiences, our language and our concepts affect the way we understand and make sense of the world, the way we decide and the way we solve problems. These different backgrounds and experiences mean that we think about things differently, but cognitive psychology suggests that the thinking process and the mechanisms of thought are the same for all of us. We all construct a representation of the world. We all selectively attend to some features at the expense of others. We all rely on memory to fill in the details and to guide the future.

So, there is not *one* way to think. There are ways of thinking. Cognitive psychology provides insight and understanding about the different ways that we process information and how we make sense of the world.

In this book, I have given you a background in psychology and cognitive science. If you read the chapters on the history of our science, you should know where these theories come from and what they say about our species' special and possibly unique ability to self-reflect. If you read the chapters on attention and perception, you know how our physiology seems to have evolved to bring structure quickly and effortlessly to an ever-changing world of sensory input. If you read the chapters on memory, you know how memory helps to stabilise our understanding of the present and make predictions about the future. If you read the chapters on reasoning and decision-making, you know how these memories and experiences usually let us make adaptive decisions and sometimes lead us to make mistakes. From perception to attention, from memory to concepts and from language to complex behaviour, our brain and mind create and recreate experience for us. We trust our sensations, our perceptions, our judgements and our decisions. We seem to be designed to trust. In fact, a systematic lack of trust in our own thoughts would be a problem. To continually mistrust our own thoughts would be pathological.

So much of what we think, see, remember and believe is a *recreation*. How can we come to terms with this?

Accept some of this uncertainty and learn to understand that your memories are not accurate. They may not reflect an accurate record of what happened to you or what you experienced. They may be missing detail. There may be gaps. There may be distortions. Instead, they usually reflect what we need to know to survive, to learn and to prosper. Our mind completes patterns so that we can react adaptively and predict correctly. There may be occasional intrusions and exaggerations. Your memory and thinking may not be accurate, but thinking is adaptive to new situations. The stretching of the truth is what allows us to generalise to new situations. The twisting of memory is what allows us to predict new features and new things. And that's what thinking is all about: learning to adapt and behave, learning to decide and solve problems. Thinking is what we do. And an understanding of thinking and behaviour is an understanding of ourselves.

References

Anderson, R. C. & Pichert, J. W. (1978). Recall of previously unrecallable information following a shift in perspective. *Journal of Verbal Learning and Verbal Behavior, 17*(1), 1–12.

Arcaro, M. J., Thaler, L., Quinlan, D. J., Monaco, S., Khan, S., Valyear, K. F., ... Culham, J. C. (2019). Psychophysical and neuroimaging responses to moving stimuli in a patient with the Riddoch phenomenon due to bilateral visual cortex lesions. *Neuropsychologia, 128*, 150–165.

Arnott, S. R., Thaler, L., Milne, J. L., Kish, D. & Goodale, M. A. (2013). Shape-specific activation of occipital cortex in an early blind echolocation expert. *Neuropsychologia, 51*(5), 938–949.

Baddeley, A. D. & Hitch, G. (1974). Working Memory. In G. H. Bower (Ed.), *Psychology of Learning and Motivation* (Vol. 8, pp. 47–89). Academic Press.

Baron, J. (2008). *Thinking and Deciding 4th ed.* New York: Cambridge University Press.

Bartlett, F. C. (1932). *Remembering: A Study in Experimental and Social Psychology.* New York, NY, US: Cambridge University Press.

Baumeister, R. F. (2014). Self-regulation, ego depletion, and inhibition. *Neuropsychologia, 65*, 313–319.

Baumeister, R. F., Bratslavsky, E., Muraven, M. & Tice, D. M. (1998). Ego depletion: is the active self a limited resource? *Journal of Personality and Social Psychology, 74*(5), 1252–1265.

Bever, G. T. (1970). The cognitive basis for linguistic structures. In R. Hayes (Ed.), *Cognition and language development* (pp. 227–360). Wiley.

Boroditsky, L., Fuhrman, O. & McCormick, K. (2011). Do English and Mandarin speakers think about time differently? *Cognition, 118*(1), 123–129.

Brooks, L. R. (1967). The suppression of visualization by reading. *The Quarterly Journal of Experimental Psychology, 19*(4), 289–299.

Collins, A. M. & Quillian, M. R. (1969). Retrieval time from semantic memory. *Journal of Verbal Learning and Verbal Behavior, 8*(2), 240–247.

Craik, F. I. M. & Tulving, E. (1975). Depth of processing and the retention of words in episodic memory. *Journal of Experimental Psychology. General, 104*(3), 268–294.

References

Dekker, S., Lee, N. C., Howard-Jones, P. & Jolles, J. (2012). Neuromyths in Education: Prevalence and Predictors of Misconceptions among Teachers. *Frontiers in Psychology*, 3, 429.

Evans, J. S. B. T. (2003). In two minds: dual-process accounts of reasoning. *Trends in Cognitive Sciences*, 7(10), 454–459.

Gable, P. & Harmon-Jones, E. (2010). The motivational dimensional model of affect: Implications for breadth of attention, memory, and cognitive categorisation. *Cognition and Emotion*, 24(2), 322–337.

Garrison, K. E., Finley, A. J. & Schmeichel, B. J. (2018). Ego Depletion Reduces Attention Control: Evidence From Two High-Powered Preregistered Experiments. *Personality & Social Psychology Bulletin*, 45(5), 728–739.

Gasper, K. & Clore, G. L. (2002). Attending to the big picture: mood and global versus local processing of visual information. *Psychological Science*, 13(1), 34–40.

Gigerenzer, G., Gaissmaier, W., Kurz-Milcke, E., Schwartz, L. M. & Woloshin, S. (2007). Helping Doctors and Patients Make Sense of Health Statistics. *Psychological Science in the Public Interest*, Vol. 8, pp. 53–96.

Goldszmidt, M., Minda, J. P. & Bordage, G. (2013). Developing a unified list of physicians' reasoning tasks during clinical encounters. *Academic Medicine: Journal of the Association of American Medical Colleges*, 88(3), 390–397.

Goodman, N. (1983). *Fact, Fiction, and Forecast*. Harvard University Press.

Hagger, M. S., Chatzisarantis, N. L. D., Alberts, H., Anggono, C. O., Batailler, C., Birt, A. R., ... Howe, M. L. (2016). A Multilab Preregistered Replication of the Ego-Depletion Effect. *Perspectives on Psychological Science: A Journal of the Association for Psychological Science*, 11(4), 546–573.

Hebb, D. O. (1949). *The Organization of Behavior: A Neuropsychological Theory*. J. Wiley; Chapman & Hall.

Heider, E. R. (1972). Universals in color naming and memory. *Journal of Experimental Psychology*, 93(1), 10–20.

Hirstein, W., & Ramachandran, V. S. (1997). Capgras syndrome: a novel probe for understanding the neural representation of the identity and familiarity of persons. *Proceedings. Biological Sciences / The Royal Society*, 264(1380), 437–444.

Hockett, C. F. (1960). The origin of speech. *Scientific American*, 203, 89–96.

Holmes, E. A., O'Connor, R. C., Perry, V. H., Tracey, I., Wessely, S., Arseneault, L., ... Bullmore, E. (2020). Multidisciplinary research priorities for the COVID-19 pandemic: a call for action for mental health science. *The Lancet. Psychiatry*, 7(6), 547–560.

Hornsby, A. N., Evans, T., Riefer, P. S., Prior, R. & Love, B. C. (2019). Conceptual organization is revealed by consumer activity patterns. *Computational Brain & Behavior*, 1–12.

James, W. (1890). *The Principles of Psychology*. New York, NY: Henry Holt and Company.

Kahneman, D. (2011). *Thinking, Fast and Slow*. Macmillan.

Kahneman, D., Knetsch, J. L. & Thaler, R. H. (1991). Anomalies: The Endowment Effect, Loss Aversion, and Status Quo Bias. *The Journal of Economic Perspectives: A Journal of the American Economic Association*, 5(1), 193–206.

Kahneman, D. & Tversky, A. (1979). Prospect Theory: An Analysis of Decisions under Risk. *Econometrica: Journal of the Econometric Society*, 47, 278.

Lakoff, G. & Johnson, M. (2008). *Metaphors We Live By*. University of Chicago Press.

López, G. (2019). *New Zealand's gun laws, explained. Vox.*

Malt, B. C., Sloman, S. A., Gennari, S., Shi, M. & Wang, Y. (1999). Knowing versus Naming: Similarity and the Linguistic Categorization of Artifacts. *Journal of Memory and Language,* Vol. 40, pp. 230–262.

Marr, D. (1982). *Vision: A Computational Investigation into the Human Representation and Processing of Visual Information.* San Fransisco: WH Freeman.

Merriam, E. P., Thase, M. E., Haas, G. L., Keshavan, M. S. & Sweeney, J. A. (1999). Prefrontal cortical dysfunction in depression determined by Wisconsin Card Sorting Test performance. *The American Journal of Psychiatry, 156*(5), 780–782.

Mischel, W., Ebbesen, E. B. & Zeiss, A. R. (1972). Cognitive and attentional mechanisms in delay of gratification. *Journal of Personality and Social Psychology, 21*(2), 204–218.

Nadler, R. T., Rabi, R. & Minda, J. P. (2010). Better mood and better performance learning rule-described categories is enhanced by positive mood. *Psychological Science, 21*(12), 1770–1776.

Newell, A., Simon, H. A. & Others. (1972). *Human Problem Solving* (Vol. 104). Prentice-Hall Englewood Cliffs, NJ.

Núñez, R., Allen, M., Gao, R., Miller Rigoli, C., Relaford-Doyle, J. & Semenuks, A. (2019). What happened to cognitive science? *Nature Human Behaviour, 3*(8), 782–791.

Oberauer, K. (2009). *Design for a working memory.* In B. H. Ross (Ed.), *The psychology of learning and motivation: Vol. 51. The psychology of learning and motivation* (pp. 45–100). Elsevier Academic Press.

Ogawa, S., Lee, T. M., Kay, A. R. & Tank, D. W. (1990). Brain magnetic resonance imaging with contrast dependent on blood oxygenation. *Proceedings of the National Academy of Sciences of the United States of America, 87*(24), 9868–9872.

Osherson, D. N., Smith, E. E., Wilkie, O., López, A. & Shafir, E. (1990). Category-based induction. *Psychological Review, 97*(2), 185–200.

Owen, A. M. & Coleman, M. R. (2008). Detecting awareness in the vegetative state. *Annals of the New York Academy of Sciences, 1129,* 130–138.

Quine, W. V. (1969). Natural Kinds. In N. Rescher (Ed.), *Essays in Honor of Carl G. Hempel: A Tribute on the Occasion of his Sixty-Fifth Birthday* (pp. 5–23). Dordrecht: Springer Netherlands.

Rips, L. J. (1989). Similarity, typicality, and categorization. *Similarity and Analogical Reasoning, 2159.*

Ritchie, S. J., Cox, S. R., Shen, X., Lombardo, M. V., Reus, L. M., Alloza, C., … Deary, I. J. (2018). Sex Differences in the Adult Human Brain: Evidence from 5216 UK Biobank Participants. *Cerebral Cortex , 28*(8), 2959–2975.

Roediger, H. L. & McDermott, K. B. (1995). Creating false memories: Remembering words not presented in lists. *Journal of Experimental Psychology. Learning, Memory, and Cognition, 21*(4), 803–814.

Rosch, E. & Mervis, C. B. (1975). Family resemblances: Studies in the internal structure of categories. *Cognitive Psychology, 7*(4), 573–605.

Schacter, D. L. (1999). The seven sins of memory. Insights from psychology and cognitive neuroscience. *The American Psychologist, 54*(3), 182–203.

Schwartz, B. (2004). *The paradox of choice: Why more is less.* HarperCollins Publishers.

References

Shafir, E. B., Smith, E. E. & Osherson, D. N. (1990). Typicality and reasoning fallacies. *Memory & Cognition*, Vol. 18, pp. 229–239.

Shepard, R. N. (1987). Toward a universal law of generalization for psychological science. *Science, 237*(4820), 1317–1323.

Skinner, B. F. (1957). *Verbal Behavior*. Acton, MA: Copley Publishing Group.

Sloman, S. A. (1996). The empirical case for two systems of reasoning. *Psychological Bulletin, 119*(1), 3.

Sloman, S. A. & Lagnado, D. (2005). The problem of induction. *The Cambridge Handbook of Thinking and Reasoning*, 95–116.

Sripada, C., Kessler, D. & Jonides, J. (2014). Methylphenidate blocks effort-induced depletion of regulatory control in healthy volunteers. *Psychological Science, 25*(6), 1227–1234.

Stothart, C., Mitchum, A. & Yehnert, C. (2015). The attentional cost of receiving a cell phone notification. *Journal of Experimental Psychology. Human Perception and Performance, 41*(4), 893–897.

Thaler, R. (1985). Mental Accounting and Consumer Choice. *Marketing Science, 4*(3), 199–214.

Thaler, R. H. & Sunstein, C. R. (2009). *Nudge: Improving decisions about health, wealth, and happiness*. Penguin.

Tulving, E. (1972). *Episodic and semantic memory*. In E. Tulving & W. Donaldson, *Organization of memory*. Academic Press.

Tulving, E. (2002). Episodic memory: from mind to brain. *Annual Review of Psychology, 53*, 1–25.

Tversky, A. & Kahneman, D. (1974). Judgment under Uncertainty: Heuristics and Biases. *Science, 185*(4157), 1124–1131.

Tversky, A. & Kahneman, D. (1981). The framing of decisions and the psychology of choice. *Science, 211*(4481), 453–458.

Tversky, A. & Kahneman, D. (1983). Extensional versus intuitive reasoning: The conjunction fallacy in probability judgment. *Psychological Review*. Retrieved from http://psycnet.apa.org/record/1984-03110-001

Vohs, K. D., Glass, B. D., Maddox, W. T. & Markman, A. B. (2011). Ego Depletion Is Not Just Fatigue: Evidence From a Total Sleep Deprivation Experiment. *Social Psychological and Personality Science, 2*(2), 166–173.

Voss, J. L., Bridge, D. J., Cohen, N. J. & Walker, J. A. (2017). A Closer Look at the Hippocampus and Memory. *Trends in Cognitive Sciences, 21*(8), 577–588.

Ward, A. F., Duke, K., Gneezy, A. & Bos, M. W. (2017). Brain Drain: The Mere Presence of One's Own Smartphone Reduces Available Cognitive Capacity. *Journal of the Association for Consumer Research 2*(2), 140–154.

Wason, P. C. (1960). On the Failure to Eliminate Hypotheses in a Conceptual Task. *The Quarterly Journal of Experimental Psychology, 12*(3), 129–140.

Whorf, B. L. (1956). *Language, thought, and reality: selected writings*. Technology Press of Massachusetts Institute of Technology: Cambridge, Mass.

Index

Index

357

Index